Birds And Other Beasts

R. H. PEAKE

Birds and other Beast

This book is written to provide information and motivation to readers. Its purpose is not to render any type of psychological, legal, or professional advice of any kind. The content is the sole opinion and expression of the author, and not necessarily that of the publisher.

Copyright © 2020 by Richard H. Peake

All rights reserved. No part of this book may be reproduced, transmitted, or distributed in any form by any means, including, but not limited to, recording, photocopying, or taking screenshots of parts of the book, without prior written permission from the author or the publisher. Brief quotations for noncommercial purposes, such as book reviews, permitted by Fair Use of the U.S. Copyright Law, are allowed without written permissions, as long as such quotations do not cause damage to the book's commercial value. For permissions, write to the publisher, whose address is stated below.

Printed in the United States of America.

ISBN 978-1-953150-35-6 (Paperback)
ISBN 978-1-953150-36-3 (Digital)

Lettra Press books may be ordered through booksellers or by contacting:

Lettra Press LLC
30 N Gould St. Suite 4753
Sheridan, WY 82801, USA
1 307-200-3414 | info@lettrapress.com
www.lettrapress.com

Also by R. H. Peake:

Wings Across ...
Poems for Terence
Birds of the Virginia Cumberlands

Contents

Acknowledgements 15
Dedication 17
Preface 19
Shapes of Beauty: Richard
Peake's Wings Across … 23

Hiking Down Straight Fork 29
    I On the Ridge 29
    II Big Oak Branch 31
    III White Rock Branch 33
    IV Church Rock Branch 34
    V Devil's Fork 35
    VI Water Gap 36
    Celebrations 39
    For Robert Lowell: January 1978 39
    Greek Gifts 40
    Residue 40
    Sails from the Orient 42
    Inebriate 43
    Taps 44
    Panther Knob 45
    Reverie 45

- Light-dark Experiment 47
- Peregrine 49
- Impassive Gazer 50
- Virginia Razorbill: Museum No. 4021 51
- Transition 52
- Harlequins 52
- Cottonwoods 53
- Final Solution 54
- Binoculated Snipe 55
- Sky Tracks 55
- Capital Snowbirds 56
- Bound by That Music 57
- Litmus Flight 57
- Harsh Gossamer: Netting Birds 58
- Sumo 59
- Prelude to Flight 60
- Neotropic Migrant 60
- Galveston Beach Birdwalk 61
- No Redbreast Herald 63
- Our Lady of the Hawks 64
- Snowy Owl in Virginia 65
- Trapping Light 66
- Death, How Can You Be Proud? 66

Birding the Rio Grande 69
- I On the Road 69
- II Brownsville 72
- III Santa Ana 74
- IV Falcon Dam 77
- V The Road Back 80

Adaptations 85
    Spring...85
    September Heat 85
    No Modern 86
    Apple Harvest 87
    Snow Music..87
    Soft Winter Song 88
    Child Roland's Dirge........................89
    Snow Crows 89
    Migration Paths 90
    Fall Flight 91
    Ice Bird 93
    It's Spring Again 94
    Unicorn 94
    Flower Garden..................................97
    Early Autumn 100
    Curtain 101
    Sing, Katydid 101
    Memory.. 102
    Jockeying Freeways....................... 103
    Jack-o-Lanterns 103
    Soul's Morning............................... 104
    Gentian Blue 104
    A Child's Dream 106
    Silly Little Sun................................ 107
    New Year's Eve 108
    Summer Crowd 108
    Old Raven 109
    Winter Moonlight 110
    Summer Seaside 111
    Ash Wednesday 111

    September 112
    Last Day of Summer 113
    Snowflakes 114
    Comic Fall... 114
    Spring Rack....................................... 115

Birding with the Bard.............................. 117
    I. To Verona 117
    II. The Big D 120
    III Arizona Peaks and Deserts............ 122
    IV Utah Arts..................................... 129
    V Colorado High 131
    VI Homeward Bound........................ 137

With and Without Love 139
    Malt, Milton, and Mary Jane 139
    Morning Song.................................... 140
    Night Winds 141
    Another Opinion 142
    Chiropractor's Dream........................ 142
    Bloodline... 143
    Silver Linings 144
    Retreat 145
    Foxtrot 146
    The Catch 147
    Yamaha Use....................................... 147
    Creon, I Have Met You...................... 148
    Needed: Pharoah's Daughter 149
    Ackerman's Laugh............................. 149
    911: Final Call 150

The Door-to-Door Show 153
    Reel I: Sampling 153
    Reel II: Pounding Pavement 155
    Reel III: Westward Ho! 157
    Reel IV: No Angels There 160
    Reel V: San Diego 163
    VI: Heading East 165

Historical Places and Perspectives 169
    Urbania 169
    Subway Rites 170
    Mound Builders 170
    Black Mountain: 1930 171
    Georgia Southern in Norfolk, 1950 171
    Recessional 172
    Jamestown, 1958 173
    Opposable Thumbs 174
    Caesar's Head, S. C., 1960 174
    Dragstrip 175
    At Arlington, 1965 175
    Myths for Space 176
    Dateline D. C., Prime Time:
    March 9, 1977 177
    Lowell Seminar, 1978 178
    Dream Trip 178
    Triassics 179
    Totentanz 180
    Primal Question 181
    Modern Technology 181
    Embarkation 182
    Paleolithic Ax 182

Children of Daedalus 183
Old Architects 183
Jilted 184
Thanet Isle, 449 A.D. 185
Anthropology Lecture 185
Dead Scholarship 187
Yeats' Statues: a Commentary 187
Intelligent Designer 189

Surreal Songs 191
The Maker 191
Kinetics 192
Hedonist at the Madhouse 192
Killing Time 194
49er: De Gustibus … 194
Sails and Indigo 195
Flambeau 195
Road Race 196
Liftoff for Mars 196
War Games 197
It's A Jungle Out There 197
Trips 198
Man's Buddy 198
Sharing 199
Sans Sense 199
Useful Catbriar 200
Cold Tracks 201
Lessons Unlearned 201
Family Tree 202
Deer Sighting 203
Idealists 203

Urban Olympics 203
A Final Solution? 204
Dental Evolution 205
One-way Trip 205
Old Wives' Tales 206
Campfire 207
Desert Blooms 207
Ostriches 208
Circe's Lover 208
Paternity 209
The Music Contest: Retrospective 210
Pilgrimage 211

Environmental Hues and Blues 213
Winter Fare 213
Passion and the Desert 213
Yard Care 214
Peccadilloes 215
Links 216
A Substitution 216
Cumberland Gap 217
Grandfather Mt. Salamander Hunt 218
Homage to Ruskin Freer: Naturalist 218
Evening Movement 219
Modernity 220
Prologue 221
Farm Lands Remembered 221
Wetlands in the Mind 222
Voyager 223
Environmental Aesthete 224
October Evening 225

Seedtime 226
Who Has Title? 226
Intersections on the Inland Waterway ... 227
Ode to a Power Shortage 228
Take It Off! Strip! 229
Silver Lawns 230
Namibian Feasts:
If You Want to Save Some Wildlife 231

Internal Blues 233
No Second Flights 233
Retrospection 234
Satellite Map 234
A World of Afternoon 235
After Parting 235
Ragged Tread 236
Intentions 237
Cover-up 237
Requiem for Nick 238
Seekers 239
Heat Wave Past 239
Winter Mulch 239
Whispers 240
Growth 241
The End of Youth 241
Debriefing 242
Philemon to Baucis 242
Bus Trip 243
Command Performance 244
Afternoon Mail 244
Fire Escape 245

The Game Is Fixed 246
My First Trip to Chicago. 1956................ 246
Thermostat Control.................................... 248

Comic Blues 251
Cheers!.. 251
Cakes and Ale.. 251
A Consideration... 252
Street Scene... 252
Vamp.. 253
Clancy to the Poet....................................... 253
Snake-eyes... 254
Adamite Orchardists.................................. 255
Indigestion 255
Politically Incorrect Advice 255
A Fact of Life 256
Busy Old Fool and Friend......................... 257
A Flamboyant's Birthday.......................... 258
Eddie Guest, I Know You 258
You Can't Hike in Kruger,
and the Zulu Are on Strike 259
Adaptation 260
Ben Ezra's Fraud.. 261

Acknowledgements

*Cumberland, Georgia Review, Impetus,
Jimsonweed, Snowy Egret,
University of Virginia Magazine,*
Vision Books, *Wind*

Dedication

To Catherine Mahony and John Mack Clarke,
who have encouraged me in my writing,
and Martha, my wife, who sustained me

"Some shape of beauty moves away the pall
From our dark spirits"
John Keats

"Men are held here
Within a mighty tide swept
onward toward a final sea"
James Still

"No living man will see again the virgin giant
hardwoods"
Aldo Leopold

Preface

My serious attempts to write poetry began when I was an undergraduate at the University of Virginia. A collection of my early poetry was awarded the Mary Cummings Eudy poetry award by the English faculty and led to my becoming the poetry editor of *The University of Virginia Magazine*. As a young faculty member at Clemson University I was fortunate to place some poems in *Impetus* together with Hollis Summers and John Ciardi. Some of my early poems such as *Greek Gifts*; *Malt, Milton, and Mary Jane*; *A Substitution*; *Inebriate*; *Cottonwoods*; and *Peregrine* appear here without much change. Others have been worked and reworked. Over the years I have continued to write other poetry. *East Beach Birdwalk* and *Ben Ezra's Fraud*.

    More than a decade ago John Mack Clarke persuaded me to allow Vision Books to publish some of my poems under the title of *Wings Across* A few years later he published some more of my poems in a chapbook entitled *Poems for Terence*. I am indebted to John for insisting that I publish my poems, the majority of them for the first time, although a number of them had appeared in journals. Since the publication of the book and chapbook by Vision Books, I have made little effort to publish further poetry, although I have read some of the unpublished poems from time to time. Through the years I have received support and encouragement from Catherine and Jack Mahony, who have read some of these poems in manuscript. Many of these poems— especially sections of *The Door to Door Show* and *Birding for the Bard*—were well received at my public readings, and I was encouraged to publish them. Excerpts from these have appeared in *Jimsonweed*. As people have expressed a wish to experience *The Door to Door Show* in its entirety, this long poem recently has appeared for the first time in *Jimsonweed*. I hope this volume will gain a wider audience for this poem and the other poems presented here.

Included in this volume are the poems from my previous volumes of poetry as well as the introduction to *Wings Across ...* written by John Lang, whose critical judgments of that volume I accept without reservation. His comments can be applied to many of the other poems in this volume as well, I believe, although some of these poems place more emphasis upon human nature than do the poems that Lang assessed. Nevertheless, I do not think that there could be a better introduction to my poetry than Lang's.

To paraphrase Keats, the beauty of imagery drawn from the natural world has always prompted me to wish to drink deeply from the spring of life. I hope these poems prompt readers to see themselves as part of a web of life that startles us with its complexity but offers us a sense of being part of a journey through universal order. Though *Birding the Rio Grande, The Door to Door Show,* and *Birding for the Bard* are all long poems using the journey motif that organizes *Hiking Down Straight Fork,* the first two of these poems use a loose blank verse rather than free verse, and the third intersperses some free verse within loose blank verse. *The Door to Door Show* was my first long poem developed in loose blank verse stanzas using the journey motif. It and those that followed owe a great deal to the example of Robert Lowell's later poetry, which I studied in depth during a seminar at Rice University with Monroe Spears in 1978. It was during this seminar that I composed the main portion of *Birding the Rio Grande,* which I completed the following year.

—Richard Peake

Shapes of Beauty: Richard Peake's Wings Across ...

This first volume of poems by long-time Southwest Virginia resident Richard Peake provides cause for celebration. In its careful, loving attention to the natural world, *Wings Across ...* follows the advice given over a century and a half ago by Ralph Waldo Emerson in his famous essay *Nature*: "wise men ... fasten words again to visible things." This Richard Peake does, participating in a long tradition of American nature poetry that began with Anne Bradstreet's *Contemplations* and continues in our own day.

An amateur ornithologist and experienced bird watcher, Peake revels in what fellow poet Jeff Daniel Marion has called the "miracles of the air." Birds appear in poem after poem in this collection, most notably in the book's longest single work, *Birding the Rio Grande*, a five-part poem some twenty pages in length. Across these pages soar Bewick's wrens, Botteri's sparrows, brown jays, paraques, golden-cheeked warblers, chachalacas, and a myriad of other birds. But *Birding the Rio Grande* does not simply catalogue their existence; it also explores humanity's relationship to nature and the poet's relationships to his father and his son, the latter of whom accompanies the poet on this journey of discovery. The poem gains depth by using the archetypal journey motif and by drawing upon humanity's ancient fascination with flight. Moreover, the natural world the poet portrays is both beautiful and fragile, both resilient and vulnerable.

The splendor of oleander hides oil tanks and U-Totem stores, and cliff swallows have learned to nest under bridges. Yet "sugar cane and concrete/ have eaten huisache bushes and mesquite." Like Robert Frost in *The Oven Bird*, Peake often sees about him "diminished things." His poems must arise, in large part, out of the desire to preserve and to praise nature's endangered beauty. His song, like that of the oriole he hears, incorporates "the sum of a world's scattered forms saved." It is "an anthem against ... forgetfulness."

From this journey south, the poet and his son return with minds and spirits refreshed, with "wings across our thoughts," the phrase that gives this book its title. Immersion in nature nourishes the human imagination, for "vireos feed our minds as they feed flesh," the poet writes. The experiences this poem recounts, in its loose blank verse lines, aid the reader's recovery of nature as a resource and instill a right regard for human limitations, a healthy humility in the presence of a world we did not and cannot make. The poem's closing line mirrors the contrast between the human and the avian realms both thematically and structurally, with its two initial trochees in an otherwise iambic line: "Heavy, earthbound, men soar as best they can."

Natural objects and features of nature predominate in the other four sections of *Wings Across ...* as well. The opening section, for instance, entitled *Hiking Down Straight Fork*, recalls A. R. Ammons' dictum that "A Poem is a Walk." Here, as in *Birding the Rio Grande*, Peake focuses on water as a fundamental natural element, essential to sustaining life. Four of the poem's six parts take their titles from the names of branches flowing into Straight Fork. Like many Southern writers, Peake expresses a love of place, especially of natural landscapes, that is complemented by a sense of history, both human and natural. *Hiking down Straight Fork* beneath "a sky as blue as Wedgewood," the poet observes a soaring red-tailed hawk and notes that the same species floated above generations of Cherokee and, more recently, above lumbermen and coal-company surveyors. Though his eye is that of the naturalist, this poet is always conscious of the region's human history as well. Amidst the bulldozer's "spoor of spoil" darkening the fork, he urges his readers to recall and respect nature's grandeur.

Yet Peake's attitude toward nature, it should be emphasized, is not that of the sentimentalist. Section II of *Wings Across ...* consists of three poems grouped under the general title *Wild Things*. Whereas Emerson heard nature thundering the Ten Commandments, Peake presents the reader, in *Winter Fare*, with a Darwinian struggle for survival, one creature feeding another "as form gives way to form." For all its Edenic qualities, nature's order is built on blood. Preying—not praying—is its vital principle, as is also evident in the poems *Peregrine, Harlequins,* and *Impassive Gazer.* The last of these, a poem reminiscent of Emerson's *Brahma*, invokes the Hindu god Shiva, who "feed[s] the roots of changing form." For Peake, however, the destruction evident in nature is part of a larger creative process that breeds life, just as Shiva is the god not only of destruction but of reproduction. Thus, even such dark poems as *Peregrine* and *Harlequins* occur in the section of *Wings Across ...* entitled *Celebrations.* Similarly, the book's concluding section, Adaptations from the German, traces a seasonal cycle that begins and ends with spring (though summer poems are notably absent, poems of fall and winter predominating instead). In this final section poems such as *Migration Paths* and *Fall Flight* reinforce the distance between humanity and such natural phenomena as birds. At the same time, through its epigraph from Columbus' diary,

Fall Flight reminds the reader of nature's diminishment over centuries of human abuse. We in the late twentieth century need to recover the explorer's sense of awe, Peake suggests, and he underscores this idea by raising questions in the poem's second stanza that resemble those posed to Job out of the whirlwind, questions meant to reveal to Job his place in a universe beyond human making.

Although death is inevitable, the poet accepts that fate in nature's design, however impersonal that design may be. In fact, in *Spring Rack*, the book's final poem, Peake envisions himself dead amidst what he calls "the revelry of grave." Instead of lamenting his demise, he welcomes the transformation death brings. Each of the poem's five stanzas ends with a reference to the laurel that grows from his decaying corpse. Here is a poet, then, willing to forgo the traditional laurel wreath as an emblem of poetic achievement for the sake of the living laurel.

Lest anyone assume that Richard Peake is a poet of only one mood or subject or one literary form, I should add that he works skillfully in a variety of modes: loose blank verse, rhymed traditional forms, free verse (occasionally rhymed), and syllabics. In addition to the many poems in which nature is his principal subject, readers will find in this finely crafted volume love poems, meditative poems, portraits of people, and the moving *For Robert Lowell: January, 1978*, one of the book's best. This book abounds in beautiful lines and images: "The black-necked waders cry in their wet fields," for example, and "skies the white-faced ibis soars." Such lines embody, in Fred Chappell's phrase, "the eye's joy."

What Peake says about sighting a rare green kingfisher in *Birding the Rio Grande* can be said as well about the poems themselves: "Delight follows discovery."

—John Lang
Emory & Henry College

Hiking Down Straight Fork

I On the Ridge

Topping the ridge that fronts the Cumberlands
talking dam
talking flood
talking wilderness,
words scurrying the air like chipmunks leap
the leaves and trunks of our advent
as we announce our jumping off on journey.

In the rubbled clear cut the juncos mill
while hikers circle restless for their plunge
down the road
to intersect with Big Oak Branch
where poplar grow tall
in the floodplain
of imagination's sunlight enclave carved
by the men who chopped oak down to size.

Whose size?
Theirs?
Or generations' yet to come?

On a way was a trail before a wagon road
we drove up Powell Mountain from Back Valley
under a sky as blue as Wedgewood china.

Uplifting a hawk the Cherokee named
eagle with the red tail
who hovers high
on thermals holds himself aloft above

a noisy group who sense his majesty
and yell and gape and point their praise
as the autocrat surveys his two square miles
of mice and meals
marking their paths
while our boots stumble by with awkwardness:
Circle upon circle he draws out
ascending—
upward, the sure-winged god
whistles his triumph.

Our collected aspirations move forth
hiking down Straight Fork to Carter's Store.
We follow the road,
the trail,
the railroad bed
and feel the wilderness envelop us
though every step uncovers artifacts

of hunt,
and dig,
and lumbermen who built
their narrow gauge
through the gigantic oak, hickory
maple to render yard-wide boards
from Boone's forest

where the whitetail waved
its flag at the Shawnee and Cherokee
and settler,
indifferent in aplomb.

II Big Oak Branch

Our hunting group finds the spoor
with quick eyes
that lead us as noses led old *habilis*...

Like a beacon in carefully raked sea
of reddish dirt a scent trap waits to track
the nosy animals that come to smell.

A little farther on, some more debris
in the road—
the white tape marks the spot
for helicopter coal survey that's gone,
but not the tape—
Consol's X marks the spot.

Then jumping off,
crashing down the hillside
into the lean tulip trees sentinel
over giant rhododendron thickets:
hearing the dee-dee-dee of chickadee
we salute *Parus atricapillus*...

Yanked up from the stream salamanders
squirm out of the grasping hand,
duskies
squirm before their caudal appendages
reveal whether they are fuscus or seal.

Along the stream we pick up leaves.
Look up.
What's this, sourwood?
Well, I guess,
how can we tell these leaves?
See there?
Up at the top, up eighty feet,
those leaves shout *Tilia americana*!
Hey, so do these,
these nutlike fruits hanging
on slender stalk from narrow bract of leaf.

O, where is the flowery white of summer
when Swainson's warblers sing the cooling stream,
the stream that laves the trees with hands of mist
bathes *Dryopteris intermedia*,
bathes *Polystichum acrostichoides*
and encourages moss to cover trunks

of the fallen chestnut, oak, and maple,
covers the trash,
covers the iron,
covers the buckets
left by lumbermen on Big Oak Branch.

III White Rock Branch

On white rocks ravished by the April flood
we sit to eat our lunch.

Look at this rock
gouged by the force,
wiped clean of algae
for a new beginning
shining white
and pink and blue.
Kaleidoscopic sun
dances across the water
running clear
enough to drink
but fierce to gouge white rock.
Rock movement here too slow for us to feel,
but we can hear the woodpecker tap wood,
see the water crash the stillness of the pool,
and listen to the laughter of the crows
who cry our hike,
yet we pass unrecognized
granitic strength that molds in silence.

IV Church Rock Branch

Just below the Church Rock we stand and wait
at Lane Camp Branch beside the yellow birch
at prayer—
forced to kneel over the Straight Fork
the birches bare their roots torn from the rock,
their tenuous grip concealed in the deep dirt.

Watching eddies in the pool at Lane Camp
where its water hits the force of Straight Fork
clouding the bottom rocks:
silt pushes gray
opaque over the pool in murky forms
as seers read the sign of the bulldozer
hidden from their view.
The spoor of spoil flows
thick,
casting its dirt on Straight Fork's clarity.

Through the rhododendron of Church Rock Branch
we tread,
white water below and hemlock
along a trail that sometimes humps the hill
for slumping down the slope toward the stream
like an avalanche of trees and black spoil.

Rounding a bend,
we look over the hollow
and see an ochre wall rising sheer
above a bare bench undecorated

by yellow machines rusting in the sun:

further on
we find fifty-foot oaks
thrown into Straight Fork from the bench above.
Across the swift water of Straight Fork
a forest lingers cool and dark with hemlock.

V Devil's Fork

As we draw closer to the Devil's Fork
we see a rock face sheltering the bank,
climb the slope to stand under rocky roof
where John "Cliff" Kerns raised a big family,
digging coal and hauling it down the stream
to Ft. Blackmore.
Scratching among the rocks
we turn up the top of John's coal stove
and soles of what must have been his wife's shoes.

We scratch at John's bones as he scratched at bones
of psilophytes or grubbed up market roots.
He dug blackberry,
cohosh black or blue,
squirrel corn, mandrake, and
most prized of all,
Sang, ginseng, *Panax quinquefolium*,
Chinese folk food to give fertility.

John dug these roots to sell at Ft. Blackmore,
but used the sassafras himself for tea
and gathered dock and dandelion or mustard
greens and cress to mix with tender poke
so his wife could set a mess of salat greens.

Like the ravens nesting on the cliffs
above the Devil's Fork,
Cliff Kerns came black
from his coal shaft where he crawled a long day
and chipped at the black face
of the bare coal seam
and loaded nuggets he grubbed
into a pony cart.

VI Water Gap

Down to clapboard house deserted now
we hike the final miles of Straight Fork
and settle down to rest
in the seeding grass.
Sitting under apple trees studying the gap
we fight with yellow jackets for the cider
of last fruits
scattered on the ground and limbs.

While the cardinals call we sit and face
the hills
covered with Shawnee war paint.

After a few farewells we climb in cars
to move apart
from Kerns
and Benjy
and Boone
remembering the leafy path down Straight Fork.

Celebrations

For Robert Lowell: January 1978

Here the jackhammer jabs into the ocean

Though you are gone and I years older now
the wind and wave still burst upon black rocks
of the New England coast. They reappear
to those who come to hear the jackhammer sea
or gaze at the gaggle of geese and gulls
and ducks that swim in the winter surf.

O executor of oceanic blues,
you blast and rage against the fluid line
of eternal beach covered with flotsam
that lures the white gull lifted by the air.

Can you feel his fast heart and narrowed wings?
Black rocks remain at Biddeford Pool
to bear the wind and wave that roar the gulf
below the gull that floats the westward breeze.

Greek Gifts

Carry me lightly, carry my bed
across the threshold
into the garden, out to the hyacinths
blooming red and cold
upon the white stone lilies of the dead.

If I must speak, if I must talk
about the world's end
I will not, I cannot offer
hollow gifts to send
to bodies withered on their stalks.

Offer up a flower, offer a rose
with a thorn or two
for the all-knowing gardener
who plants the bulbs, who wipes the dew
away with the spout of his hose.

Residue

Sheen of fireflies
seen in loose flight
flaming through the dusk,
without the bugs to fire them
the vegetables and flower plants
fade away.
With them comes a sound
in an eager ear
and a throb in the chest

singing in the key of desire.
Upon this newly shaken star,
our world,
the days and nights breathe
beauty flashing light
in the evening.

Darkness follows, then morning
that sharpens with an iron file
the dull knife of longing
for soft and warm embracing skin
velvet for flight down a scented aisle
toward a key above
the gale ringing
the grinding voice of love.

Love sounded in salt cinders and mist
molders among shadows
on lips of remembrance that insist
of roamed over thighs and of fallow
lust gentling as a face bends its wind
down on another's field to blend.

A wish runs by the mind
like a brushfire
turning reason to black
smoking ash.
A pile of dust remains
and chemical analysis
reveals much calcium
and many traces of elements

the rains will turn to lettuce and mint
and coreopsis and chrysanthemum
to delight the bugs
and light the fire.

Sails from the Orient

"Those were the times," the old lady said,
"when the white-rigged ships brought incense
from Cathay to waft me a delicate murmur
of golden, silent faces politely eating mince
and sipping their unlemon-puckered tea."

"Don't mind her," said Joseph, and reached
me another, more modern social drink;
though I, polite to memory, muttered a demur
before I lifted up the proffered glass.

"The birds came every spring then,"
she told us as she continued tatting,
"I liked to think how neat they had returned
from the other hemisphere, but I knew
the summer's heat would make them somber."
"Get her," Joseph whispered with superior wit,
"the old girl has a feather fetish;
doesn't she know she banked her fire forty years ago?
Besides, she's lived with steamboats all her life,
and they don't give a damn for sailed exteriors."

She droned on about the beatific birds
of spring that had disappeared down the horizons
of those far-off festival summers
when she had bewitched men instead of thread,
about the numerous wedding cakes her mother baked,
too late, and how the bird nests
used to tumble down the chimney before Joseph
put detestable central heating in the hall
and ended dreams of sailing ships.

Inebriate

Two soft doves on straight wings
over a bush of red quince
are my love's eyes
that sit soft and pliant
above the wet shawl of her lips.

Sweet fresh apples on a high branch
under the windy dress of the sky
are my love's breasts
that quickly rise and fall
in the veined white of her skin:

and I would share the flight
and taste the sweetness of the fruit.

Two white birches side by side
bending across a dark mountain
are my love's arms

tapering from firm shoulders
over the shadow of her hair.

Warm, summer hills
rippling in a soft breeze
are my love's legs
glistening in deep twilight
before the red wine of her thighs:

and I would bend the birches
and dream upon the hills.

Taps

A wood pewee
whistles and mourns
in dusty dark.

I stare at empty sky
and listen to crickets drone—
do you wait in dusk?

At my foot a leaf moves,
red as blood that flows
from a cut on my hand—

red as blood that burns
when my dark reverie
remembers your touch.

Panther Knob

Solstice sun burns
at the painter's throat
melting its white.
At rest in a bough
of the reclining cat
a rusty blackbird
whistles kee-o-lee
from the leafless tree
over the dark back
after winter sun
in somber decline
lets the wind chill
the painter's hide.

Reverie

I am alone at midnight.

The rabbits shiver
under the snow banks,
and under the blanket
my arms ache
for the lover
whose face I cannot see.
Her hand caresses me
in dream.

Gray horses
charge on a field of cloud,

race the winds,
swift across paddocks
of salt-marsh grass,
past the pole at the far turn
of Bodie Island light,
past the slender form
thrust up from the sand.

The gallop of red stallions
careens across the clouds
formed by warming sun
to cool the beach
on which the Atlantic
throws kiss upon kiss.
I lie on the shimmering sand
without a footprint—
the sun embraces me.
The breeze tingles.
With a lover's touch
it teases emptiness.
As the gulls laugh
a brown arm blocks the sun
and arches over my head.

Bright eyes unwrap me,
their warmth
like the sun.

The screams of the gulls
echo in my mind
like voices in a dark hall.

I wander the wax myrtle maze
behind the dunes
and scavenge wood
to match my horses.

Then blue horses race
The evening sun to rest—
sea moods shift
like the geese of snow
that straggle skeins in the light.
I pluck out their down
in setting sun
and ride
on white wings.
A lover's hand
offers me a drink
from a crystal goblet
filled with wine
drawn from fountains of spice.
Her voice on the April wind
whispers my name—
only she can hear
my dream.

Light-dark Experiment

Assembled beside a pond
they break out meter sticks
to measure the biomass:

light uncovers

emerald wing
of a brown bird,

**flung against
lazuli sky,**

in plummeting earthward for a walk
ungainly with the grazing geese
beyond the gaze of workers
spreading bottles on the waters

**finding life
in the pond
dark with algae.**

Leaping fish announce
tingling ears immeasurable waves
on mists of ochre smoke and rock
uncovering grunting caterpillars—

hurling

the green-winged teal
beyond the field.

Peregrine

in gunpowder sky

aglow with sunset,

a mourning dove in straight flight
over a field of brush
where I stand hidden
watching swift movement

suddenly

broken

by the strike of talons

downtoppling on bowbent wings

stooping

on the soft feathers—
bird with the black mask

STRIKES!

(lover striding fire and feeling of spherical bust)

rowing wings
leap forth a falcon force ascending,

guy talons trailing blood.

Warlord of the upper air.

Impassive Gazer

Lone Shiva is the stony god
who contemplates both birth and death,
the change from foetus to the clod
uniting earth again with earth.

When sky blends wind and sun with bird
Shiva is the master there;
when talons topple down unheard
the falcon launches Shiva's spear.

Shiva fills the egg and womb,
watches doves and rabbits born
before he sees them to the tomb
to feed the roots of changing form.

Shiva watches hawk and hare,
the falcon and the fleeing dove;
his wraithlike gaze is clear and bare
of sentiment for strife he wove.

Virginia Razorbill: Museum No. 4021

A drowned Alcidian sailor
lies before us on the beach
half-covered by the winter sand.
Reverently we lift the body high

who stand awe-struck to find it there,
saddened that his fishing days
have ended: swimming on the waves
this beauty would have brought us joy.

Picked up from the sand, a razorbill
reveals unhappy death in youth
by his dark and half-grown beak.
That bill will cause fish no fear.

The harsh wind from the bay
little comforts us who mourn a bird
whose strangeness to this shore
dictates it lie in state as specimen.

Feathery suit of formal black
and white must grace a museum case
and gain a numbered immortality
far from the ice floes of his youth.

Flocks of scoter blacken sky above the bay
and the large gulls hover,
scavengers angrily mourning lost corpse
as they wheel above us.

Transition

Swifts stutter in autumn air,
chitter and sift insects in the gray light,
whirling through the half-knitting sky
into October chimneys.
Autumn is for apple trees
and the last bee workers
in labor on the goldenrod
of a roadside close to the riverbed
and to the season where gray clouds
hitch-hike on an asphalt sky,
bound far and high
on cold thumbs to catch a ride
with the files of geese
winging, winging, winging
in forks from the Back Bay
to the corn-stubble in the fields.

Harlequins

Caracaras
laughing at noon
bash meat of a coon
to blue flesh barrows,
tasty carrion
for Caracaras—
clownish waste barons
live on dead flesh
moved by worm harrows
dressing meat fresh

for Caracaras,
who laugh at sun's baste
warming bone marrows
to epicure taste
of Caracaras.

Cottonwoods

The willows of the desert
are the cottonwoods
who sit tall
and lean
along the stream in council
about
the asphalt of the motor road
that moves
sinuously
where buffalo stood
on the grass that fed Indian ponies.
Until
the long rifle spoke
with a finality
in the sunrise
as
the mountain man sought food.

But the cottonwood
does not seem to be
a tree
with weeping tendency,
and even if it were,

its tears
would get
short-shrift
from a dried-up stream bed
lying inimical
under
the desert sun
watching gila monsters
on the ground
and buzzards in the air.

Final Solution

Roan monkeys spout in wrath
to ancient Hindus
left in Borneo holes
with bright but dusty wraiths.

Leopards growl and dart
at jungle redwings
of Angkor black in ruin,
great, majestic, and deaf.

Gazelles run the Khan's gain
of sands to Samarkand
from the empire of Kin
while spears rust to grit.

Afghans have found their graves
on cold Himalayas;
no Afghan kings on horse
raise their swords from the ground.

Binoculated Snipe

The stolid, humble tiller of the land
stands by his fence and smokes a pipe
in silent wonderment about the band
of stalkers who binoculate his snipe.

Fearlessly they tramp his marshy bogs
chasing cows and chickens everywhere.
Without a thought of omnipresent hogs,
they stop knee deep in muck to stare.

"What are you looking at?" the farmer asks,
and shakes his head when they reply, "Your snipe."
While all the watchers reassume their masks,
the farmer looks amused and puffs his pipe.

Sky Tracks

Whorling up, a snowy mass churns
its ink-tipped arching feathered wings
above the darkened stubble in the burn
men charred for greener marsh when spring
comes grassy mantle cracking black-
ened earth while white birds beat on overhead

the bay, their color hiding wrack
they rend in winter fabric's spread
as snow geese rise to wing *en masse*
north to their tundra home once more.
Flight brings flight as birds move past
on aero tracks their ancestors wore.

Capital Snowbirds

Gallows drew a man from earth inert
jerking rope of toughest hemp
iron shod with flecks of frozen dirt:
silhouette whose collar held him limp
under the white ptarmigan, dangling hurt—
bird's flight defying puppet's eyes
whose jellied gaze seemed wise.

Lifeless eyes stared at winter air
after wheat preening golden pods
under summer's cloudy brood mares
sweating dirt into greening sod.
A corpse was roughly put to bed.
They who dropped him there have trod
the rocky ground that Judas fled.

Quick hand slash threw one thought stranger
hanging dead down onto frozen grime.
He seemed harbinger of danger,
not their brotherhood in crime.

Sanguineless was put to bed betimes
a man whose fellows watched him swing
and heard the white snow crunching.

Bound by That Music

Many youths caught in love's harsh music
remind me of that Nashville Warbler
we found on a foray in Virginia
among the northern hardwoods and spruce
and a lone tamarack where he picked a bog
he liked and set up shop to sing his song.
He didn't seem to mind that we who heard
thought him an oddity who wouldn't find
a mate. He had the confidence of youth,
for he was sounding still his ringing call
when we forsook our wilderness he owned.
Were those emphatic calls we stalked to hear
wasted? Certainly they were not meant for us
though we happened by to notice them
and reacted with a fervor kin to his
to find him so far south of normal range.
I hail youths' rock rhythms. They hurt my ears,
yet the primal urge brings all songs to birth.

Litmus Flight

A white-winged gull invaded my youth
on a spike of light from the cold, gray truth
of an April day. I braced against the gale

blowing from the east as I stood with my back
to the tidal flats along the river's track.

Against the April sky my litmus gull
turned pink to raise the ph of my soul.
Whipped by the gusting wind, it soared and told
its rippling wings to stride the roller coaster air.
I wracked my brain to try to join it there,

but my bright gull ascended into clouds
still crying its strength and grace so proudly
that my feet grew heavy with weight of man
as they sank into the river mud and sand,
earthbound, not fit to pass the acid test.

Harsh Gossamer: Netting Birds

In the afternoon I walk softly
Down the hollow and drop in place my nets
To catch unwary birds that hit the web
I've set to capture curious flight.
The tufted titmouse screams as my hand frees
Its wing. Somewhere deep within me cries
A captive answering the caught bird's desire
For feathers striding wind to fly away.
After I take them from my nylon web
I carefully place a band upon one leg.
Then I measure the bird, and let it go.

Other netters net for meat, not science.
Like giant spiders that set delicate nets
To gain a feathered meal, these netters place
Homemade webs along the paths where birds fly
Moving north or south as nature dictates.
All spider, they spread their nets for food.
They quickly eat the birds they catch, relish
the savory meat of avian breasts
without a twinge of guilt or pity.

Sumo

On the smooth, wet sand
shorebirds spatter by us like corks,
bobbing heads and running in bursts at the land,
splashing stranded water, saluting the sharks
who play beyond their surfing band.

Awestruck, on the shifting dunes
we keep time with beat of the surf
welcoming the cloud-wandering moon
to the wrestling match below the turf
of dune grass. We share the view with grebes and loons.

Flexing its muscular hide
the ocean churns up white caps
and coughs out the rinds of ships that defied
its strength. Watching waves from the shore they slap
we sense the thundering waters' stride.

Prelude to Flight

Avian eyes peer through the trees
standing under their airy road
as the breeze
strikes red and yellow birds to goad
them, urging flight to stroke
the skies above the oaks.
Their wings rift
air in a frenetic, twittering surge
as their urge lifts
them up and feathers merge
with the moonlit heights
of their long night.

Neotropic Migrant

Every spring an April fool
Announces the snow's defeat
And claims its fast retreat
Will let the rule
Of bees return upon the clover,
Winter's over.
Then orchid flowers
Shoot up under the cleats
Of vernal showers
And the sun's hot feet.
Defying a late freeze
In budding trees
A red-eyed bird
Makes himself heard.

That emphatic bird
Incessantly sings
And makes tree trunks ring
Even in mid-day heating
Until the summer's over—
Says that summer's end never
Comes while a vireo's heard,
But early frosts deny bird's
Incessant cries for summer.

Galveston Beach Birdwalk

On the right we have a Caspian Tern.
Don't confuse it with the Royal near by.
They're both big birds and new birders yearn
to turn them into Sandwiches. Who knows why.
No, Mary, I don't mean the kind you eat.
I mean the Sandwich Tern. It's smaller
and has a yellow tip to its bill, I repeat,
it's not a gull or heron. They're taller.
Look at that gray and reddish bird out there
running madly around with wings outstretched.
That's a Reddish Egret. There are a pair
of those running clowns. Look! they're catch-
ing fish by spreading out their wings to get
their prey—a sight you won't soon forget.

See those black-backed birds with orange bills
that seem to flush without provocation.
Watching them when they skim the water thrills
anybody whose soul's not on vacation.

The long lower mandible seeks out fish
as skimmers fly in careful formation
hunting with technique we think outlandish,
yet we look in wonder at their action.
It awes us to see their numbers beat air
wheeling back and forth as if on some cue
eluding us. We cannot fathom where
that Peregrine that launched this melee flew.
The hunter stooped on skimmers bringing shock
to end the noontime siesta of the flock.

There, on sand, a Piping Plover scurries
in straight line. Suddenly she stops and throws
her head back, as if she's greatly displeased
with what's ahead, as if a plover knows
it doesn't pay to move too fast and miss
a worm or other prey. Her behavior
separates her from sandpipers on this
beach, scuttling back and forth seeking favors
washed from the Gulf. Plovers and sandpipers
must contend with man and the dog he brings.
These foes harass less than developers
whose machines destroy dunes and flush wings
to places less fraught with man's intrusions,
but birds don't mind our scoping them for fun.

Hey! To your left are large shorebirds, Mary.
They wear tuxedoes and long, thick red bills
that open oysters with no shucking fee.
Oystercatchers roam oyster bars to fill
their stomachs with tasty delicacies.

It's plain their bills are utilitarian
but beautiful. Like the yellow beaks
on those Great Egrets fishing there, they can
kill. Down the beach you should see pelicans—
notice that the white ones are a third again
larger than the browns: two species whose plan
for fishing is quite different. The brown's plain
force. See that dive. The whites swim in a group.
In a circle, they herd fish in, then scoop.

Look there. Those Neotropic Cormorants
stay here throughout the year unless winter's
weather gets too harsh. These birds need no plants,
only fish to warm their belly feathers.
Double-crested Cormorants come down south
for winter. Larger, they like the ferry
landing, rougher waters, and the bay's mouth.
You tell the two apart by their throaty
patches. It's yellow on the bigger bird
and angles down straight at ninety degrees
while the Neotropic's throat looks blacker.
But either's beak can give the fish a squeeze.
Our East Beach walk won't be complete today
until we see these shags find fish to slay.

No Redbreast Herald

I saw a robin the other day. It bled
in a winter field whose sky was cold,
and geese and snow stood overhead.
I'm not like those ladies who are told

to phone the papers every spring
to tell about the robin hopping
and pulling worms on Sally's lawn.
My robins aren't absolved from ice,
they're my companions of every dawn.
Come down into a swamp woods right
below a pine-scrubbed Virginia hill
when the water's high on the tupelo
and covers the cypress knobs until
you see what I said you could. You go
see a robin chucking in a holly bush
and blackbirds sounding a rusty hinge
as sun shines down in a sudden rush
when the gray and thickened clouds impinge
on squirrels chattering about nuts crushed
to brave cold metaphor with tasty heat
for melting figures wrapped against the sleet.

Our Lady of the Hawks

There are those in our society who scoff
At chasing birds without rifle or shotgun
They laugh at binoculated scoping toffs
Who think a terrific look gives lots of fun,
That added to their lists a rare wildfowl
Is better fare than roasted ptarmigans.
Whether we wanted a hawk or owl
Myriam Moore helped us to praise and prowl.
She knew the way to make birds seem *chic*
To folk who heretofore deemed it antique
Or quaint to rise at dawn to count a rail

And search in fields odd hours for a quail.
We'll miss her when the hawks are flying through,
But in her honor we must count a few.

Snowy Owl in Virginia

My joy depended
on a misplaced waif
whose advent ended
my long wait.

There he sat before flight
in the mouse-rich field,
a huge mass of white
whose fate was sealed.

Followed, haunted, gazed
at by hordes of people—
his yellow eyes looked dazed,
the pupils widely dull.

He did not eat the voles
and rats that he could see.
He longed for the cold
and starved amid plenty.

Not yet dead, examined carefully,
a rare bird for my tally.
Ave atque vale.

Trapping Light

The bright world around shouts color to us,
but we see rainbows with distorted lens,
both fleshy lenses nature gives for focus
and tools on which our photo forms depend.

The play of light upon our world betrays
the magic way feathers can trap sunbeams,
but tensile strength of feathers makes strange plays:
what's dull can sparkle evanescent gleams

when the sun's rays strike our large eye's delight,
but then the sun shifts and a bird turns gray—
its electric blue transformed without the light.
Refracted light creates a bright display

when myriad feathers trap the sun's rays—
a nonpareil bursts in bright flame anew.
What seemed a burned out coal just yesterday
becomes an indigo bunting—bright blue.

A company of eye, and sun, and stance
determine how we move in optic dance.

Death, How Can You Be Proud?

Soaring buzzards flash a sign of death
that should remind us of Prometheus,
yet for years men have learned the shibboleth
that humbleness gives us impetus

to heaven. So we all have kept our livers
from the feasts of vultures and from Lucifer.
For as long as man has known that death
finds all us men, followed by great buzzards,
we poor mortals have cursed and spent our breath
on our hatred of night's darkness and the guard
we all must keep against defiant end.
We find a carcass round and ripe too morbid
and must approve the nausea that contends
against our thinking buzzards intrepid.

We see buzzards soaring and sailing, their wings
dipped with white, a flapping menace obsessive
like the whiteness Ishmael's great whale brings
to spite our Ahab, silence expressive
of grim origins that make us doubt that white
things will live as did the Phoenix —we dislike
to see intrinsic virtue in a dog stripped
bare by Jove's gigantic vultures. The decay
of a dog covered in black is a display
as old as the pyramids of Egypt
or the cleaners perched upon the Parsees' ground.
We may never accept a death as proud
when it smells of stinking birds and makes a loud,
discordant music on a swelling hound.

Birding the Rio Grande

I On the Road

Prelude to Texas, the swamps of the coast
stretch along the highway in bayou country
while we crane our necks to see the herons
soaring with necks tucked in and legs stretched out,
a stance that looks comic, though efficient
to my son and me, speeding intruders
thrusting over concrete stilts into Eden.

Later, on dry land we water the roadside
and whiff the gas which seeps from loosened pipes
that have ravished the entrails of the earth.

The black-necked waders cry in their wet fields.

The black francolin sings his harlequin
Beep Boop Beep Boopy Oop from a tall pine,
another interloper staking a claim.

Searching for the glories of San Jacinto
we climb into our second skin, a shell
that sends a sense of power through our nerves
as we, now godlike, start the car and set
a dynamo at hum for the inner man,
a sensuous harmony these fields don't bring.

We had this once before, in the first garden
just for a moment, but now in chariots
we gods can roam our ordered networks
that crisscross crazily on the stretching plains
of Texas, home of the steer, the lone star
and consummate passion of the motor car
that carries us easily over the trail
the wagons lumbered to the Alamo.
Turning south we leap toward the Rio Grande
under skies the white-faced ibis soars
over roads where scissor-tails catch flies
fuming the humid air. Along the sea
at Galveston a Gulf breeze burns our cheeks
as we eye pelicans from a parking lot.
Down the Peninsula we trek the sands
now conquered by the crab-like homes on stilts
and hurry on to meet the southern night
of heat, mosquitoes, and paraque
whose plaintive purrweeel is funeral dirge
for bugs. It calls us to a midnight pause
within our sleeping bags against the ground.
Awake, we slap some mosquitoes uncaught
by the paraque or the nighthawks.

Clouds dot the sky that stretches horizons,
mile after mile of sorghum in the sun
except where thin thickets hint that flatness
was not always so tame. A roadside stop,
planned for breakfast and parula warblers,
offers Turk's caps in the shade of oaks
where tropical kingbirds yell of danger
as we invade, despite the sign: *Watch For Snakes*,
though flowers crimson and gold underfoot
would tempt us to dream another Eden,
if we weren't too busy dodging cow dung
and drinking color, to consider serpents.

After we break camp under cloudless sky
we move across the plains, and I have time
to contemplate the vitality of youth
that presses us along his way. The terms
are even here. What advantages gray hairs
give fade before the heat, the youth, the fare
of this new world. A time was, I remember,
when I could face eternity that way
and didn't hug each morsel of beauty
as if it might be my last. I thought
romantically of death but didn't know
him well enough to see his vacant eye
in the return of earth's quick children,
the lily, the rose, the sun on marsh at dusk.

Still, the new bird surges through my body—
with double exultation I shout out,
my son's careful excitement reminds me

of my buoyant birds of youth, but I share
with him what Father and I shared rarely,
a oneness of heart, and mind, and soul—
a black-bellied tree duck accomplished that,
breaking across the blue, *Dendrocygna
autumnalis*, our spring though in my autumn.

His flight in the sun must count for the gamble
his parents took that he might give a race
to dragons of twentieth-century man.
I stand face to face with filial love,
whose breath is life, the sharp, the bright, the rough.

II Brownsville

We find Laguna Atascosa, the Gulf
again, and the gulf between the old world
and the new. The sugar cane and concrete
have eaten huisache bushes and mesquite
so fast they have developed indigestion:
Harlingen, Brownsville, and Port Isabel,
where oleanders, red and pink and white,
show their splendid hues in front of hovels,
oil tanks, highways, and U-Totem stores.

There everyone takes the Yankee dollar
from the gringo tourists with a curse,
for city men like some kinds of green
though they hate mesquite. At noon we stop
to hear the chachalacas laugh at us.

In the willows and brush of the refuge
the racket of chachalaca sounds
outdoes even the comical rounds
of great-tailed grackles, whose cackles'
infinite variety seems subterfuge.
Karrack! Whoever heard of clowns in black?

The lordly great-tail struts his feather stuff
upon the ground before us and his mate,
who admires the shine on his coat and voice,
the *savoir-faire* of Don Giovanni
lightening the all-brooding world she lives.
Watching her sidle, we are caught in the
yellow eye of the grackle and cavort
in glee to the tunes he chortles and cracks.

Myriad small birds leap in sudden flight;
they throw their sparks of red, blue, gold, and green
into an American sky, defying
gravity, buntings painting our heaven,
though the old runways remind us sternly
that these acres of brush once served to train
young men for flight into blossoms of flak
in machines groaning, defecating bombs
before the keen-eyed hunters killed those geese.

At least the splattering excretions
dropped by chachalacas and buntings
will fertilize the field to bring new growth,
not flowers more savage than all the weeds.
What we really seek here, though, are small, brown

little sparrows, found nowhere else with ease
in these United States. Scarcity
gives Botteri's sparrows a bit of class
we think, who brave the mid-day heat to hear
them sing. Otherwise how can we tell them:
they look so much like Cassin's sparrows.

So we drive through the brush along bayside
and listen. One sings, we stop, intently
gazing at the song. A bronzed cowbird
hears it too, and flies to look, his red eye
glaring at us and searching for the singer
fled from parasite. Later we find a song
for us, lifted beyond the cowbird's stare.

III *Santa Ana*

Technology presses against Santa Ana.
Two thousand acres is all that's left
of the brushland that set the Rio Grande
apart from the southern Texas plains—
now farms and factories call a modern dance,
circle left and sugar dough, swing your gal,
then head for easy, air-conditioned home
to match your tractor's radio, cool cab,
and filtered glass that saves you from the glare
of a fervid sun against a humid sky.

Along the roads, the paths of Santa Ana
the native birds and plants hunch up for safety.
no other place to flee, the ebony tree

spreads its old age for us and the oriole
building its pendulous nest. Whatever
is fugitive insists on loveliness.

Looking across our supper at my son,
I see my youth beleaguered by my thrust
into the warmth, the old futility
of our single mother's birthing womb
in which we seek the breath, the blood, the kiss
to meld us one with her fertility.
More practical, my boy breaks out the tent
while I daydream about the fathers' eyes
I see through him. Fathers … old men with a hoe,
you have come back to me, resurrected
in this youth who works so steadily there,
his mechanical skill containing you both,
my stubborn dad, the other dad I loved,
the best men in their best world possible.

Dad's blacksmith shop and fields of wheat and corn
are far from here, but the diminished things
about me turn my mind to losses felt
more deeply than the loss of Texas plants,
yet linked inseparably. Daddy wouldn't leave
the land. He held it for his son and daughter,
a solid gift he would not break for tax—
were they worth it, those long days from the chair
you fought our pleas to sell and live at ease?
The last hug, the last tying of our eyes
you knew would never meet again in life,
though I, unwilling to accept our fate,

said goodbye and not farewell in that last clinch.
You kept trees green. What more can a man do?

Bentsen State Park and Santa Ana
hoard a remnant of tropical forest,
a miser's pile of dew glittering green
in a morning sun, a cool retreat at noon,
a tired man's bivouac at the evening wind
that springs from southern Texas ground in gusts
of glee whenever the sun goes down.
We drove into camp at Bentsen on the wings
of this breeze that blew us to orioles:
Altimira and black-headed—hard to find
that bird—but he blew in with us and sang
in the hush when the moving trees were still,
his song the sum of a world's scattered forms
saved, an anthem against peals of forgetfulness.

In the evening quiet the doves call softly,
insistently, mourning diminished day.
We search through the mesquite and oak to find
the white-tipped dove's low-pitched ooo-whooooo
that haunts the trees around us, echoing despair.
Far away, only the hollow, long-drawn whooooo
reaches ears straining to find the source
of all that woeful beauty, that dirge
following the path of a setting sun.
Found at last, the birds show rufous underwings,
throats and foreheads fronted with white.
The light at rest on their bodies turns roseate.
When we are lost in the dark that follows,

I would give eight hours for that late bloom.
A sense of loss colors the wings of morning
before we seek the birds of Falcon Dam.
As we delay our leave from Bentsen,
a harsher note invades the uncupped ears,
a sound in keeping with the raucous note
of the chachalacas: crowing roosters,
the white-winged doves asking, "Who cooks for you?"
at us, sitting below, eating cereal
cold in the open box of convenience—

our spirits soar a little with the hawk,
bay-winged beautiful that graces our sky,
but the brown cowbird, young interloper
fed by a tiny olive sparrow
blind in love, betrays our snug earthly nest.

IV Falcon Dam

We reach our destination in the heat
and view the Rio Grande from Falcon Dam.
Man's work, high, mighty, this raptor cannot fly
With the supple grace of the peregrine
That races river birds for prey each fall.

Golden woodpeckers dig for grubs in the trees
whose painted buntings sing against concrete.
The view from this man-made mountain catches

winding waters. Shallow, still, they wait
the hand that opens gates, creates the rush
upsurging river over the bare rocks.

Now the watching heron spears with ease
the fish collected handily in pools,
his stony cups for drinking frogs and snakes.
Excitement uncontained now, we bound
down the slope to the river's edge. Low water
means we may catch the green kingfisher's act,
a dive to joy made only in clear pools
where minnows cannot elude his beak.
"What's that so swiftly cutting through the air?"

"That there? A swallow by his graceful flight
and the sheen rippling over his back."

Delight follows discovery. This bird
is the emerald we birdfishers seek.
Perched, alert, he allows us his profile,
all beak, removing our lingering doubts.
Health to the beaked, green head! To the destiny
that brought us, anxious, eager, to his fief.

Joy of the bird—we sail with swallows
coasting the river though we sit on the bank
on lookout for that other jewel
of the Rio Grande, the ringed kingfisher,
whose breast emblazoned with brick red sets fire
to our imagination for high flight
that carries us to Falcon Dam State Park,

where nonpareils sing in the mid-day heat
that sears the brush of the desert birds,
the pyrrhuloxias huddled in the shade,
pink, tinkling echoes of woodland beauty
set among the devil's tongue in yellow bloom.
Sound and color tipple eye and ear asking
how this harsh world can conjure vision up.

After noon we seek out what we are told
is the "hottest" birding spot on this river,
the Santa Margarita Ranch we reach
despite directions. From the King of Spain
the Gonzales family gained this land
by grant. Raising our binoculars
we say, "pajaros," as instructed,
only to hear, "Two dollars a person
to see the birds." Mike tells us he has come
to visit, to help Grandfather. Summer
is a holiday from Houston traffic.
In heat we trek to the riverside
to search for the only brown jays to cross
from Mexico. We find boys spearing carp.
Minute follows minute, hour tedious hour,
the boys take up their fish and leave us.
Hot and tired and jayless, we hike to car.
"No jays," is the sad report we give to Mike.
He whispers in Grandfather's ear, then says
we can come back tomorrow, on the house,
no fee. Their birds don't come with guarantee,
Mike jokes—brown jay is Uncle's C. B. handle.
Jays have a special place in the life here.

Supper at the campground raises our hopes.
The cactus wren, the braggart, yells to us,
and a scaled quail sits on a bush and screams
like a guinea hen, "Jay-ho, jay-ho,"
as though a thing infallible were planned.
In the shower house, after supper,
we take off a week of sweat and dirt
under a water head, and look for ticks,
tiny specks of brown that can balloon out
to raise the temperature higher.

We move into the evening wind gusting
camp and hike to find a paraque
as a full moon casts its samite light
over prickly pear and hides the danger there.
Lights from passing cars disturb our eye
adjusted shapes. The wind moves on our minds;
the red eyes of paraques glare at us
when we turn our flashlights on the birds;
red flares fly up and mingle with the stars.

V The Road Back

Next morning brings me a jay and a son
who gnashes his teeth at me: he didn't see
that great brown jay, big as a hawk, that stole
upon us in a flock of cousin green jays,
who called him in awhile, not long enough.
We sit in the cool shade of the river
trees and watch the kingfishers green and red

shimmer in the early light rushing up
the Rio Grande. A mourning dove circles us
who have invaded her brooding life
and sit beside her nest. Beneath the sun
a hawk rises on the heated air.

Time past for hunting jays, we must head north.
Who can live a week back, to find a brown jay?

Sadly retreating to the upland hills
we hear a verdin fussing, feeding young;
the trill of a black-throated sparrow
rings upon awakened ears: sitting on a branch
of mesquite, the stripe-faced bird eases fate.

Our idyll ended, the Rio Grande fades
to memories, bright but poignant with loss
of red-billed pigeons, rare, maroon in sun
streaking shafts across the water's morning
birds that fly away ... wings across our thoughts.

In the shade of a lookout's hill we survey
the Rio Grande. One last time. Mockingbirds
serenade us, vultures circle our sky.

The light that blears my eyes is not the sun.

We move toward the west along the lake—
it touches our path and wets the desert;
at the high bridges swallows mill the air
and line the thickened wire beside the road;

these man-made cliffs the cliff swallows make home,
their concrete crags beside the waters.

As we roll into the streets of Laredo
we find the ballad loses something
in translation to another century.
Rock doves and house sparrows fly from sidewalks
to hide in the Spanish architecture.
The streets of Laredo lack romance now,
we spy tourists and five-dollar bills—
the mind travels uphill in motorcars.
A hundred miles of Texas plain lies between
Laredo and Antonio, and green grass
cutting across the range. The hill country
breaks the flatland at San Antonio,
and here we stop for one last look at birds
before we travel east. Friedrich Park
glows as we climb up in the evening sun
amid the noise of Bewick's wrens
who scold our search through pine and oak to find
golden-cheeked warblers and black-capped vireos.
Rufous-crowned sparrows gurgle as we watch
fledgling vireos feed in the dying sun.
Vireos feed our minds as they feed flesh.

Before we get into our car, a lark
sparrow throws his twilight song at the dusk.
As we thrust against the dark of the plains
toward the garish glare of Houston,
this prayer for peaceful sleep softens the air
of sunset behind us. As we move east

gaudy scissor-tailed flycatchers glow
in the departing light and bull-bats
flit the sky. The traffic rushes by us—

to unknown destinies. With bright lamps
cars catapult against our dreams of Eden.
The birds of southern Texas settle
to rest on the mesquite in our minds,
burned as our second gasoline, regret.

Soon my son and I must part—our birds fly
in skies with him, on a sleek super jet;
our birds stay here with me, watching the sky,
wondering where their brothers, sisters go
and whether flight will bring them safe again
together. The sparrows of the airport chirp
in tune to the rock dove's banal coo,
dismal surrogates for the birds of sun
we saw when we birded the Rio Grande.

They do their best, their very homely best,
these foreign feathers that are home here now,
creatures of a feathered beauty that shames
the metallic birds we fly to serve our need.
Heavy, earthbound, men soar as best they can.

Adaptations

Spring

Over the river fly
the seagulls tossed by wind.
The ducks chatter and lie
still on tbe riverbank.

No clouds have climbed higher
than these moments in March.
The seagulls reel and cry
of sunlight on the heart.

September Heat

Hot woman, guard your glowing touch.
Though your heat shrivels anxious grass,
the corn loves your red wine and the clutch
of grasshoppers who sing your praise.

The prairie lark that climbs the blue
applauds your sunny, glittering dawns

and morning glories. Purple dews
are shouts of triumph being born.

Your sparkling nights offer cool balm,
a liquor that helps green grass revive.
Snails wander dew-wet evening calm
but don't notice the star-dappled skies.

Your hot laughter hides fear of frost.
Your sticky toad tongue catches flies.
Your salamander heart is crossed
with browning growth and frequent fires.

No Modern

Easier than Oberon
can mount with locusts and larks
over forest and lake
the mole makes lawns dark.

The myth of Oberon
today excites harsh jibes.
His golden horses limp
whenever he drives.

The old god's harness clinks
as he battles wind
whose force chafes
the rocky harbor, my friend.

Apple Harvest

While ladders poke the apple trees
and offer young men to the evening;
they stand sky-high and fondle honey
heaven, full of golden mist.

A radiant morning follows bees
as sun ferments—summer vines
trail through blackberry, thorned memory,
apple scent, nut stain, cider, wine.

Time for a girl and a place
to cling. Ground without girls burns.
Lovers push, thrust face to face,
hug in the path, laugh their desire.

Maple leaves smile red decline,
draw glance and lip to lip.
All things that live demand to find
warm love whose wildfire will consume.

The longed-for paradise holds distress,
halts our bodies and seizes grief—
consummate summer, summer we confess
has bees that sting from their concealment.

Snow Music

I hear music in cold fields.
The winter throws the colorful snow

against the sun to create rainbows.
It plays a music I can feel.

The snowflakes tap a few bars
of brightly-colored blizzard—
of icy flowers, shining stars:
roses bloom out of crystals.

Gentle sounds strike the trees,
where snowroses wax and wane
as snowbirds call quiet refrains
that put discord of heart at ease.

Soft Winter Song

The hunter tightens the noose
but the fox gives him the slip;
the cold wind solders the loose
ice shut over eels and carp.

From arctic cold, the thrushes flee.
Their voices fall gentle on snow;
no knife cuts into earth's sleep,
the mole raises no sores.

Larvae become quiet and wise,
hang in wait for April wind;
udders of pregnant sheep rise,
and the goat's beard lengthens.

Child Roland's Dirge

Horns of summer fall silent in acres of death,
cloud after cloud floats into dark
and circling forests grown thin
like mourners who follow a hearse.

Terrified, the brown fields moan the storm
that strips the poplars to white towers
and sweeps across the barren:
gray huts of a village huddled below.

The tents of autumn corn spread
far away, stretch into desert.
Countless cities stand forgotten, empty,
and nobody moves through the streets.

Only ravens, shadows of the night,
fly under the thickening clouds in the rain.
Alone in the wind, like dark omens of sleep,
these black thoughts flee through wind-swept hours.

Snow Crows

Black crows on a white field:
the sight clutches my heart.
Snow flurries encircle the world.
Solemn crows sit and wait.

Magic birds from ancient times,
they come to visit us

in gallows' clothes. The crime
they wear echoes the old curse.

A new spirit walks the forest
but crows do not see it move
beside them in the whiteness
or hear it sound in winter woods.

Old crows cry out raw and hollow
unhappy tales of ancient worlds;
they squat in the dark, and snow
swirls chill as night uncurls.

Migration Paths

These tracks have enchanted men a million years,
roadways moving birds high through the sky.
Like a ball of blue in a mist, the earth
turns as wings measure themselves against air.

The weight of men denies them these paths
refusing to lift cloven hooves, or paws.
No dust stirs there, no leaves fall from the bough.
If gods are there, they have no hair on their brows.

From such a track in air above their heads
men see bird shadows break about their feet.
The beat of thin wings uncovers the light
they see flicker in the lightning of the strokes.

Men have created bright metaphor;
the words have battered them with rough force.
Men's superficial magic chases birds
along the endless, invisible arch:

One end of the bridge rests in the frost
where sphagnum oozes and the owl circles
and pumps its feathered tail red from the rust
gathered in the stain of liquid iron.

At the other end the fish and insects hide,
and the greedy flock to this food and eat.
The noise of a moving stream resounds
where the avian road sinks in the sedge.

A journey brings birds to their winter feed
which heavy frogs and snakes or bears can strike
only if they fly to catch the soaring dream.
If only men had wings to take the paths of birds.

Fall Flight

"Todas la noche oyeron pasar pajaros"
Columbus' *Diary*

"We heard birds in steady flight all night"—
myriad wings rustle through our dreams
and hide Orion from watchers on the roof
who turn their scopes north to the horizon,
and high southward without knowledge or wisdom.

What calls birds to the skies on August nights,
who tells the thrushes and snipe to set forth
from forests and marshes; caged birds flutter.
What instincts move this nighttime armada?
Jet liners delay take off tonight;
(All radar sets flicker in confusion)
pilots have to go to bed or get drunk,
controllers gaze from towers with blind eyes
into their cones of light that shed no light—

the birds blip; all night long across the screen
the cormorant carries food in its pouch
for its young; golden eagles seek rabbits,
a red falcon clutches a quiet rodent,
while a griffon, long thought extinct, hunts lion
in the minds of watchers fed by radar.

Unseen birds grow larger, fiercer than those seen.
All night long the small songbirds and shorebirds
flash tiny silhouettes across the moon;
for hundreds, thousands of miles their ceaseless
wings embrace the air and trouble dreams.

We study flight and learn migration paths,
but we will never lose Columbus' awe.
What threatens the skies in these latitudes
And what awaits us when these wings have flown?

Ice Bird

A bird glides
through the vastness of space
on unseen lines—

with the grace
of a redbird
on a crockery face—

frozen and unheard
in the azure,
its wings undisturbed,

this flight is secure:
over blue ice
the dance is pure.

II

The beam in flight
to the radio set
pierces ears, a light

whose sound has fled
like a tone from struck glass
echoes in the head.

A silver tone can flash
over calm ocean
where hikers cannot pass.

The play of frost upon
the ice cuts with a harp
whose high note numbs

It's Spring Again

Two eggs, a loaf, a hat, and a hound—
in spring sky the white wool
streams out, free
in a heaven without background.

So, uncultured friend, speak freely,
and tell me what I had.
Tell me what I was. A plough?
I had what I have now.
How do I know what I am? Fears?
So stupid, after sixty years.

A croaking raven,
I fly over my doubts and away.
The wind kidnapped my hound.
The eggs, the bread, the cognac
taste good today.
And look there: my old hat
makes little men and little paws.

Unicorn

A night wind
tugs at the heart;

it murmurs the bright art
of spring.

In twilight treetops peacocks preen
blue, golden, green;
and gray primates tumble,
grin and snarl, scuffle
in the undergrowth.

A tiger crouches at wait,
lies low, watches, flashes its claws
as bewildered deer dart
through the forest
westward to the sea.

The unicorn's glee—zestfully
his hooves battle waves—deftly
he runs with the hart
and hunter.

Overhead dark cranes fly
riddles in Arabian sky.
They offer us mirage with sinking sun—
scarlet apples, yellow pears,
slices of ripe melon,
peaches, oranges, bunches of grapes.

Forts of amethyst,
enchanted castles of topaz
and vermillion glow—
rose mists flow

over a dark bay.
Unicorn day—
his hooves swirl
the soundless dust that furls
around minarets and gravestones.
They lie hushed
under a singing moon.
A night wind
tugs at the heart;
moon whispers the bright art
of spring.

Unicorn eyes
embrace cold stars
and cities bleached for ghosts
who hear the desert owl,
the jackal howl,
hyena laugh.
Beside old mud walls
under date palms
a bell calls
when a white dromedary
lifts a calm head.

Move unicorn, move
unweighted feet
toward your fate;
your gaze glitters
like serpent eyes entranced
by the flute to untwine and dance.

The unicorn
runs in the night;
he carries high horn
with grace in his middle brow
and sheds soft light
on uncovered face
and naked breasts
of the woman who waits
to end his race.

She greets
him with her eyes
and murmuring lips, fountain—
night winds
tug at the heart;
they whisper the bright art
of spring.

Flower Garden

Nothing else moves, no bird, no breeze,
a silver leaf flutters and whirls.
Come with me, girl,
the dew-breathing grass lies cool
like magic on my feverish foot.
Look around at our green sea.

The great crested woodpeckers
plunder nuts and crack them.
Look there, they've drummed on that log.
the broken shells litter the ground,

but the birds are not calling now,
the trees are quiet.

Red apples flame upon the grass.
I try to shake off gold-green plums—
I didn't forget you, wait for me, please.
I must ease my hunger with fruit
that fell from those empty trees.
What about you?

Can't you see butterfly peacocks
light on the rotten apple peel.
Hold me. We will sip like the butterflies
who draw the juice from fruit and reel—
their colors blind.

A mourning cloak spreads its velvet,
purple golden fringe, on that rose—
stoop there and smell.
Yesterday it was only a bud;
the smell is sweet and pungent now.
Its odor charms.
Soon night will close upon the bloom,
this flower open now will die.
You know the hummingbirds and honeybees
must drink from out of flower mugs.
Do bees dream?

You will dream if you drink bees' milk,
sweet and creamy gold elixir.
Loveliness spins the handsome bees

that whirl, furious, hot-tempered skins
into swarms.
Why do you look at me and laugh?
Do my eyes sparkle with your wine?
Will you laugh at my pulsing hand on your lips—
morning wind—soft swaying stalks,
will we drink in the sun?

Made of bronze, a tiny frog calls
from water, and a mermaid stirs,
freezes our blood, woman whose hair flows
in the pond, enchanting me.

We turn our faces from the bathers.
If you give me your flowing hair
to ripple between my fingers,
will you mock me?
Like the growth on the river's bend,
tender, clasping camouflage—
will you laugh?

The wind whispers your flight—
my fondling hands press soft plumage.
Are you the swan the great god of the ancients took
to create boy twins and Helen?
Are you Helen?

My eager hands smooth the feathers
as broad wings flash high symmetry.
I understand ... I dreamed ...
a garden breathes when sun and rain

enflame the blooms for honey bees
and hummingbirds who taste the nectar.

Early Autumn

An insect sings
to the evening star.
The apples ripen
to their cores.

The treetops tire
from heavy fruit.
Fog climbs up high
to pasture.

Berries push themselves
against berries;
the bumblebee tells
of aster honey.

A pear on the tree
matures and swells;
another pear
ripens and falls free.

I suck my cheeks full
of juice to dine
on the fruit
that offers first wine.

Curtain

When the mists creep
Like a night cloak
Over mountain rim
Through chestnut oaks

Then sounds grow dim
As the earth stills
For evening sleep.
It closes petals.

Sing, Katydid

A katydid sings,
an unknown artist
of the evening
storms the heart.
Memories of a kiss
rush in the window
where you rest an elbow
and heavy thoughts.

You are there alone,
and the shadow image of our earth,
the moon, comes secretly
through the trees.
That weird crone,
the night, full of mirth
and stars, takes little care
of your despair ...

An eye is open still
though it is deep night.
On these shaking ramparts
time passes.
Damp air makes you shiver,
but soon it will be light.
Then the solitary watch departs.
Life passes.

Memory

Years ago, each night
I was a pine tree
That clutched a cliff
To clasp the moon tight
With white hands.
Above the abyss
I bent in dance
And grabbed the mist
And clouds, my fleeting toys,
Above the steep rocks.
I feel no sorrow
Or wild delight, no joy,
When ecstasy fades
And pine bark sleeps
In deep shadow.

Jockeying Freeways

You jerk! You cut me off. You're real cool.
Get a nag in Tarquinia, driving fool!
In those tombs it is still the era of men,
The skies belong to birds, the seas to fish,
The earth to the bull and the noble horse.
Time here ticks off the folly of insects.
Or, have you covered up some secret news
From the tombs to the century of cars?
Then turn around, while the tunes of the dead
Still fend away the swarms of maenad bugs
And the scorn of apocalyptic hornets.
Drive down to the cities occupied
By pious age and guard the beautiful,
The paintings, until the horn of morning
Calls you to ride on a red Etruscan steed.
Too slow for you, though, you're such a sport.
Freeways demand more modern transport.

Jack-o-Lanterns

We've thought a long time
About jack-o-lanterns.
Sprightly aunts, they remind
Me of family concerns.

Between these clowns and me
There are close family traits
Few others have. Everyone can see
Us dance in the swirl of our race.

Yet I go where no pathways go,
Few men, in swaths of mist;
And I have watched winds blow
Closed the eyelids of dead artists.

Soul's Morning

An outcry from sleep
Creates a sense of deep
Sadness and cold, icy cleats
As winds retreat down dark streets
And circling stars undress
Their comeliness
In green rivers, silver alleys.
Towers gleam at new trees
As sun winks through
And drinks up purple dew.
Bluets spring out new dresses.
A quiet drunkenness
Moves a gliding boat.
Sun lovers gloat.
Songbirds pale
As lovers laugh and whirl
In children's garden worlds.
Primroses break the veil.

Gentian Blue

Here early snow sobs
Into moss mats.

Here knees throb
On steep paths.

The wind sings
A chill evening
To a child
Who asks why.

Gentian fall
Dares the first frost.
The flowers sprawl
Against rocks.

Sea drinks the small spring
Unseen in mountain thickets.
Only a salmon, dying,
Colors distress.

A gnarled fir
Still stands.
A mountain goat hangs
To cliff like a burr.

An old mountain,
An old strife,
The gentian
Comes under cold knife.

Ice and talus
Sleep and wait.
Only windy gusts
Awake.

A he-goat hurls
A bell on the wind.
A small girl
Grins.

A Child's Dream

Drunk with poppy juice and sad thrush note,
In silence, a bearded man touched woman.
His mother bore her child in a white moon
Under shadows of old maple trees.

In the soft window dark, old heirlooms
Lie in decay, old dreams of love can fall
Within father's moonlit ancestral vault,
Forgetful of cool breezes, green branches.

In dark days of his years, desolate childhood,
The boy imagined fishes swim cool streams
And threw himself down under horses hooves
Racing through black night as his stars seized dreams.

He held the freezing hand of his mother
In the dusk as he crossed leaf-littered yards
When autumn brought a frail corpse to bed
In that dark room where he opened his eyes.

He was just a small bird in leafless trees.
In the shadow womb of the ancient trees

Bells rang at the stillness of his father
Who climbed down the winding stair to sleep.

He held the bony hand of the old man.
They went at dark to the falling walls
Of the city: a black overcoat dragged
The child to the maples; the ghosts approached.

Grasping green stuff of summer, how softly
The garden fell apart in autumn brown,
In scents of sadness from the old maples
When the boy heard the thrush and angels.

Silly Little Sun

My little sun ... when you rise in the sky
you are no bigger than a sparrow's head.
Before the cuckoo has three times told why,
the clouds with rain will wet our bed.
We are not supposed to see the blue dome
then. My brother, the wild tongue of the hound,
sports a blue flower, but who takes him home
for comfort under skies filled with summer sounds?
What good comes from golden stars if buttercups
cannot counter the infertile machine's shove
and no maidens kneel to pick blooms up
to question if their lovers truly love?
What is the use of forget-me-not dares
if the girls ignore the wonders of sleep,
staring their dreams away like nightmares
until their visions turn and maidens weep

away despair. My brain, too, has fled
from the warning calls of the rain-bird nation:
"You are no bigger than a sparrow's head
when you rise in the sky, my little sun!

New Year's Eve

Through the turnip hole a mouse jumps
As wind breaks dry leaves from the oaks.
The hollow turnip gapes at stumps
While gusts whip the freezing dovecote.

Stout farm horses dress in long hair
Like wild beasts. The year comes and goes.
The oak falls, its cry lost in air
Like shouts over icy ocean floes.

Summer Crowd

Man-high wheat stands for harvest
Under a pale moon in twilight;
The grain gloats at nightmare forest
Who listens in the dark night

The sickle mows the mass of gold,
The mower bathes in his motion,
The brown hands stretch and fold.
As feet edge forward mice run.

Who leads the cutter through the fields?

Why does wheat whisper in the furrows?
While the ears hear the wind rattle
And wail, the heads bend and bow.
Those that dare to speak mutter.

The billows of grain that strut the land
Rail at the lightning that kills them,
But they are stretched out by its hand
And their wide field becomes sodden.

Where have the golden days all gone?

The crowd whispers we are here still.
No one believes happy days will end
Though the sickle sounds a high peril.
Each stalk hears and trembles and bends.

One night the sickle strikes this place
Under bright stars and batwings.
The nightmare makes a cunning face
At the stalks. He knows what frost brings.

Who ponders where the guys have gone?

Old Raven

I saw a bird
Long years ago, heard
His croaks near old gray towers

In London.
He told that travel sours.
It had undone
Him.
He sat black and hopeless.
Legend says
He sat there speechless
Hour after hour
And watched old kings' treasure
In the tower
Year after year.
Today I know
The meaning of that dream,
Its power.
I have felt regret
And seen that bird again.
I have seen inside the tower
At night and know
What wealth he watched.
Lying worn, I heard
Him say his truth,
"Lost, youth."

Winter Moonlight

I watch February moon lie
On clear heaven, turquoise-blue.
In winter grasses, yellow fire,
Sheep move and rest and chew.

The ram aches for beauty's ewe.

Wool gleams star-cleansed coral.
I know the words to raise the moon:
I am in Paradise before the Fall.

Summer Seaside

A napping cat is not undone
by hot stones or noise from a house
whose clock and stereo sounds
prompt noonday dreams as sea soaks sun.

Bathers on the beach burn asleep,
have lost the world and do not care
that smells of sea or sand they wear
sprint with the sea wind from the deep.

Empty space creates cloud cities
while wavering heat waves massage
skies' freckled gull and tern montage
whose muted cries cause no unease.

Across the summer heat birds fly
and light on the fishermen's stakes.
Dark thunderheads begin to make,
yet threatening clouds escape hot eyes.

Ash Wednesday

Yesterday I wore rainbow scarves
Through peacock, singing worlds.

Today my desire starves,
Its peacock plumes all furled.

One eye looks back that way
The other looks on dearth.
Quickly, suddenly, flower earth
Shows me no blooms today.

The moon is yellow, flat, untrue.
The skies are cold and blue,
My wood lies under icy stars,
Its peacocks all departed.

September

You, who lean on the fence in front of phlox and oaks
(splattered by rainstorm and the scent of wild beasts)
who like to walk the stubble and talk to the old folks
gathering juicy autumn apple feasts,
you breathe on the fields smoking with heated foggy cloaks.
You lack the will that snow and winter bring
to yell, "Hey, you're wasting time,"
at the vines climbing up. The summer workers sing;
you restrain your voice. Workers chime,
"You're short and fat and heavy-set,
rotten pumpkin beside your shoe, wet
fungus without face, slimy growth and grime."

Climber from the plains, flames from a final moon
swelling out fever and fruit, drooping face already
dark—
knave or fool or baptizer,
idiot of summer, clattering echo, death rattle, crooner
of morning song of glaciers,
brush-cutter, nut-cracker, adage-eater, shark.
In front of you lie snow, high silence, and barren
space without plants: your long arm reaches there,
yet you lean over the fence as beetles throng among
the greens—
all life eats life—spiders and field mice and hares
and autumn leaves are chewed by winter scenes.

Last Day of Summer

Day, you end the summer games.
Your signs all fall apart
And set off forest flames,
Autumn hues that tug the heart.

After you, pictures turn pale
And snatch at the cold times.
Water glistens like flat ale
Even though waters are wide.

You must learn how to fight
And battle to slow your flight,
To which the swarms, the crowds,
The armies of ice bring shrouds.

Snowflakes

Slowly turning green limbs white,
nothing speaks with greater softness:
like lovers whispering low at night,
who sleep together in their strangeness
and in the morning's brightening light
find an odd earth has built for their delight
a nest full of celestial whiteness.

Comic Fall

The clusters of grapes
ripen clear through;
frost transforms green shapes
with dreamlike hues
as winds weave light.

Leaf after leaf falls
to dust the lawn;
a white hand mauls
the woodland thorns
as trees strip bare.

The hounds and hunters cry
what life has breath.
The dead year's drunken hires
bequeath what's left
with absurd hope.

Laughing, the first snow
gleams on mountainsides.

Bird flocks dull its glow
but cannot hide
the silver course of death.

Spring Rack

A fish in a brown sea of cognac,
my corpse swims with white belly
firm, intact. From the drunken sack
on my skeleton, laurel grows.

I have reached the revelry of grave,
a drunken ghost among spring roots
whose drinkers dance into my mound—
from out of my heart the laurel grows.

Burgundy red streams from my pores,
the restless vines spring out of me
while mustard seed sprout in my ears—
from out of my chest the laurel grows.

Blue eyes lengthen on the larkspurs,
tips of buffalo grass grow pale
as hungry moles grub in my stomach—
from out of my groin the laurel grows.

Drunken riot resounds the earth,
the worms that feast my body's core
have broken down the bones to dust—
from out of my loam the laurel grows.

Birding with the Bard

I. To Verona

Journeying west, we hope to see the bard;
the birds, and a country called America—
like a sea journey, though earth's hard
core grounds our quest for celestial liquor.

We are almost self-contained, pared to bone;
in our truck we stow food and sleeping gear
so that we can see wild lands red-bone raw
between those nights we soak up Shakespeare.

Trip to give art and nature time to heal
the pain in my side and the ache in my head.
Ventures in art and nature's art I feel
will weigh against desire to wake the dead
and tell them that they cannot tie my heart
to a rack of memories, marriage woes

of lovers of an old form grown apart
visited to the second generation:
a roadside lark lifts its voice in a pose
suggesting a songbird's celebration.

Tonight we look into curio dens
offered charitably to men and women
who try to stay cool in the evening heat.
We read the welcome banners on the street.

Tonight we shall see the *Two Gentlemen*,
tomorrow *Romeo and Juliet*—

we mix birds and bard in a westward trek
begun in Elizabethan 'Bama
where red elephants vie with high drama
for those whose entertainment reeks of sex.

The arts can flourish under a big tent
before their patrons have learned discontent.

Bald spot floodlighted on Romeo's head
does demand disbelief over dismay
to think him young and yet so hairless.
We can accept our youthful heroes bald
with power of strong imagination.

We make far greater imaginary leaps
imagining touchdown runs heroic.
Consider, what is an instant replay
but aesthetics shown the athletic way?

There was a time I remember yet
when ardent nightingales sang all night
for me, young Romeo, and Juliet.

Having spent an evening and matinee
with Alabama's answer to Stratford
-on-Avon, we turn our trucking bed and board
westward toward the Father of Waters.

At Vicksburg we pause to honor the dead
on the bluff above old father's voyagers.
My son cannot easily understand
why the brochures offer little quarter
to the Confederate dead. Bloody hands
still wave bloody shirts in federal words.

The monuments seem to echo the curse
of separate blood. "Why do the Yankees
have all the big memorials?" he asks.

"The victors get the tombs and eulogies."

Memories of feuding Capulets
and Montagues fresh in our mind's eye
we watch the river below us; we vet,
surrounded by monuments to fratricide—
marble wavers in bright light, clownish
like the balding spot our Romeo showed
under the merciless spotlight's beams.
Union, Confederate sorrows made dreams
sweet with grief for novelists and movies.

Warblers and wrens chant hymns in the thickets,
celebrate blue and gray with equal zeal:
the cuckoo's elegiac rain dirge peals
against the light engulfing all hard tombs
in the heat... chants for passionate debris.

Flight
moves through beatitudes
and eons—

thin-winged
meteor gray bird slips
through air thick with afternoon heat waves,
A Mississippi kite
seeking snakes,
swoops down and eyes us—
flips over the ridge
to soar in blue-ringed clouds above the river,
flies into thunderheads—
melding with sky
like my first Mississippi kites on the Savannah
that fled
into high ceiling, free:
This bird we share expands in time beyond my ken.

II. The Big D

Across high asphalt of a river bridge
the land lies flat nine hundred miles or more
before the rocks rise up from desert brush
to foretell the Rocky Mountain ridges.

Open air, damp commodious tile marks
water spots unlike those pioneers sought
to slake their greater thirst on westward treks.
Cactus wrens and western kingbirds drink here,
too, and they and the Crissal Thrashers yell
as gold wings flash in manzanita hell
to set brush on fire and warm tile glaciers.

Along the Dallas freeways we seek theater—
Shakespeare's shrew will play the Texas sky.
Nighthawks skim above us, cowboy boots
promenade before us, giving Mantua
a sagebrush flavor to match the stucco
stage that Katharina treads in cowgirl style:
so we are come abroad to see a world
as hot as Petruchio's Italy—
the dry air cools to aid the wooing fight.

O slanderous world! these Texas artists
are apt as those of London to hold our gaze
in thrall. No curtains mark this thespian feat:
we share Hortensio's shock the shrew is tamed
with Dallas girls; they walk the aisles, discreet.

Interstate, desert brush, and blue sky
patched with fluffy clouds flash by, flash by—
the hot Texas plains stretch out before us
a long carpet with lots of brush—design
broken here and there with the apparatus
of oil wells, some resting, some on line,
pumping rhythmically, black seesaws

working to raise their quota of dollars.
In the distance the shine of El Paso
gleams below the ridge, beckons...Rockies Ho!

Birds circle the ridges (swooping swallows)
towering above—red and blue, rainbows
to egg on heated, dry-mouthed travelers
toward cool, clear waters of old stories.

"What's that blue streak that flew through the air?"
We stop and look in trees under the bridge
to find a male Blue Grosbeak preening there.
Looking at the high peaks ahead, I think
of a fellow who laughed at the ridges
of eastern mountains. "You call those mountains?
Why those are just foothills. Now the Rockies,
they rise up and let you know they're there.
I tell you we've got lots of hills back home
bigger'n these ant bumps covered with hair."

III Arizona Peaks and Deserts

When I see them rise in clear desert air
these peaks do dwarf tree-clad Appalachians.
Those forested hills we left behind
are filled with warblers now, exuding birds
red, blue, and yellow—dots of feathers glow
against a green sea of rainforest leaves.

Here we see a different wilderness.
The Chiricahuas loom ahead. Cactus

flowers paint hot, rocky hillsides yellow
and red and green. We ride through hot canyons
where ocotillos raise their spindly arms
and climb steep, winding roads eight thousand feet
to Rustler Park, where we camp for a week.
Events do not fulfill advance billing.
No rustlers are here in the seventies.
We bird and await Shakespeare in Utah.

The little meadow below the campground
is an enchanting spot that turns blue
with iris in May and delphiniums
in August, both attracting hummingbirds
broad-tailed and rufous—jewels stun birders'
eager eyes that follow their flaming hues.

We wake at early dawn and look for birds
before we bathe with water from the one
campground spigot (what is termed a sponge bath
is not immersion but it does cut dust).
Red-shafted flickers and western bluebirds
and red-faced warblers decorate our morning.
Robins, boldly ubiquitous, make us
feel at home, but huge ponderosa pines
alive with Olive and Grace's warblers
let us know this is a new world entered.
When we descend into Cave Creek Canyon,
we find elf owls and gray-breasted jays
to complement dainty painted redstarts
in trees where truly elegant trogons
nest in old Arizona sycamores.

The time comes to power our yellow truck
toward Patagonia, the roadside stop
where many reports of marvelous birds
originate: rare Rose-throated Becards,
Beardless Tyrannulets, Thick-billed Kingbirds.
Outsiders are fenced out from the dude ranch
but do not have to chance trespass. Becards
have hung their nest just beside the highway
where every passer-by can take a look.
This doesn't include the dude who creeps through
wayside in a green Jaguar in first gear,
one hand on the wheel, the other flipping
through Jim Lane's guide to Arizona birds.
It is rare to see birders such as he
looking as if they've just stepped from a page
of Jim Bean's catalog. Three-thirty-three—
the desert birds are quiet except for that gem,
the lazuli bunting serenading me.

An old school bus sits in the wayside park.
It's still there when the evening turns to dark.

Old man in a bus
a long way from Tennessee
talks about hard times
and his woman
who matches his social security check
and grey hair
with hers.
The heat of Patagonia
never forces his lady to eject

from their yellow earthcraft.
"Four days we've been here,"
he says
and leaves us to imagine
what sanitation on that bus
is like.

Fleeing heat, we seek Madera Canyon
by way of Ramsey Canyon hummingbirds.
We pay our fee and sit in the lawn chairs
casually placed among a myriad
of feeders that hang from many limbs
and dare birds to visit for public stares.
Magnificent green hummingbirds attend
two feeders. A blue-throated stakes a claim
determinedly to one before my face.

Madera Canyon brings adventures
with odd owls and sparrows and goshawks.
Beside a motored camper we hear
owls whose bleat confuses tenderfeet.

The western screech owl repeats his howl
long enough for us to figure out
he's not just another misplaced easterner.

Later we hike up a steep slope to find
the whiskered owls that inhabit here.
In full moonlight we hear the sound and peer
into the trees where the *boo boo boo...boo*
seems to originate. Our luck has changed,

who missed the wary Flammulated Owls
along the Chiricahua Barfoot Trail,
flushed out before us in our daytime treks
and rained away from owl hunts at night.

In daylight we climb up Madera slopes
to find a brown-throated wren singing out
to lure us up and up. When we head down
we thrill to see swift flight of a goshawk
descending before us. At the trail's end
we hear screams of distress along the stream
and run to find a young Cooper's Hawk
in the clutches of the larger goshawk.
We interfere with the ecosystem
and the Cooper's Hawk flies back to the nest.

In the bright sunshine of Floreeda Gulch
we look among the brush, grass, and cacti
for rare Rufous-winged Sparrows. They are there
and flush long enough to make us happy.
Their color's drab, but rareness paints them fair.

A Tucson gem is the Desert Museum
where we can see rare cacti and rare birds
before we ascend Mt. Lemon high
above the sweltering saguaro heat.
Here cool streams and trees amaze us—this place
cools us the way the U-Totem ice cools
our food. Mountain breezes delight our faces.
The guidebook agrees this air blows cooler—

it's cold as ice a few miles higher up
when we climb from saguaro cactus flats
into the pinyon oaks as dusk brightens
city lights thousands of feet below us.
The breeze at eight thousand feet tightens
our skin a bit. We put on coats, move out,
and camp. We light our labor with starlight.

Next morning we hike out to see the sights.
We revel in profusion of flowers
and the belligerence of a Broad-tailed
Hummingbird who has staked an angry claim
to flowery roadside. He hurls his power
at those who test. He plays no favorites
between us, broad-tails, or golden eagle
soaring overhead, brown god of the heights,
who returns our eager gaze with sharp eyes.

We take the lift up to the mountain top
and loaf under the ponderosa pines

while clownish Tassel-eared Squirrels define
us as dangerous. The mother removes
her young from their hiding place to find
other, safer spots. Squirrels take babies
from forest spire to forest spire of fir.

Before we descend the slope on the lift
a western tanager bares bright beauty,
black and yellow and red, flaring in sky.
Eyes spot the Golden Eagle soaring high

above. Adrenalin excites the mind
and binoculars sharpen our weak eyes
as the brown-winged god circles overhead,
climbs, then stoops with grace below the ridge.

When we arrive at Aravaipa Creek
the desert heat seems dry and bearable.
Along the creek we feel humidity.
The heat clutches cactus and jeers at us.
We slump under a cottonwood whose girth
announces centuries of sunny earth
and life-giving water drawn from the creek.

Bushtits
and vermillion flycatchers flit
from twig to twig—
they wear a gaudy rig
that seems to turn up the heat.
The black hawk screams,
"If you can't stand the heat
get out of my canyon."
He does not soar at mid-day
but stays over the stream
and laughs at me—in his cool
and rippling pool
I collapse
and gulp for breath
beside fast waters.

IV Utah Arts

Quickly black thunderheads roll from the west
announcing time for hasty departure.
We are not fast enough. From summer heat
the landscape changes—the jagged lightning
preludes the deluge and slippery roads.
We slide to stop and wait under eerie
night in day, wondering where dryness went,
thinking forward to the heath where Macbeth
would hear the Utah witches prophesy.

Later we head north to Cedar City
through the pines of northern Arizona
whose top peaks testify to rainfall high
enough to grow trees three hundred feet tall.
Brian's Head lies on his great rocky bier
in Utah. The Indian chief has lain prone
to the skies for centuries. Recently,
his repose has been taken in vain
by sightseers who desecrate his bones.
Brian weeps rocks into valleys below
while wind-torn pines stoop in his cheeks
to form his hairline. Hordes of red finches,
pepper red, Cassin's scatter like dandruff
among the trees. From his eyebrows we peer
into the red rocks—pimples of his face
offer primitive carvings in sunburst
exploding when clouds shift their embrace
over arena of the Great Spirit.
The great chief weeps and sighs at the smog
blowing eastward from the Kaiparowits.

The ghosts of the Anasazi dogs
howl their lament from the Kaibab plateau
to sandstone winds and waters carved to form
great obelisks to awe the Navajo.

Measure for Measure marks our first foray
to Cedar City. The motel gives us
a chance to wash grime of the trip away
before we seek the solace of the bard.
Unfortunately married Claudio,
lovely Isabel, lusting Angelo
rival Brian's red finches and blue grouse.
I feel pain in my side begin to grow
as Mistress Overdone's lust brings down the house.

Next day at Lava Point among aspen
the sky-blue mountain bluebird sings to us
while eagles soar above yellow-headed
blackbirds outsinging pond frogs and bees.
The black-chinned sparrows in the chaparral
of lower slopes brighten evening pall
but the sudden breeze makes quaking aspens
rustle under a full moon and bend
in the firelight. *Unglaublich!* German speech
floats to us—bits on the wind in the trees
from flickering campfires neighboring ours
lending a mildly cosmopolitan touch
to wildness. Dry branches spark in the heart.

The Green River country must have caught
its name from imagination, or envy green

at humid Colorado ground above.

Again we pursue Shakespeare—in Cedar
City lights. This night noble Macbeth bleeds
for us—tainted by a bad director
who cannot decide whether the lady
is for burning hell or just a happy
housewife helping her husband get ahead.

V Colorado High

The land we move quickly through next day
is land where solitary ravens play
a desperate game. The careful descent
to Colorado could cause some heartburn
for those drivers who must go slow or burn
out brakes. Runaway truck ramps astonish
us—one safety spot ends in sheer rock wall
and another leaps out over a gorge.
Ravens gather
at rabbit kills on the road.
Like Odin's dead
rabbit corpses
wear raven shrouds
at dusk.
Most perching birds
depend on husks
of camouflage
or minute size
to hide their fear.
Ravens

find black
a hue that suits
their oracular act.
Grand Mesa rises against the parched plain,
whose beauty we have to hurry through
with only a stop or two to complain
about green grass and water rising up
over a parched land where even a Raven
marauding must carry his own grub.
In Mesa Verde foothills magpies cavort
in trees that shade the cattle from the sun.
Arlechino—
O fortunate eye
to behold these magpies
on the foothills of Mesa Verde
cavorting on fence lines
and trees and cow pods,
a context that defines
the rustic loafing at ease,
raucous, loud, chatter ironic—

unlike those city 'pies
we later see high
over the golden arches,
black-billed magpies
burnt by setting sun—they fly
above MacDonald's.
I take their silence for disdain
of Boulder's traffic.
The scene is hardly sylvan—
a metropolitan frieze

in which aboriginal clowns
act solemn.
At dark the yellow-bellied marmots sing
their siren song on rocks of Loveland Pass.
Their melodious squeals and grunts ring
dulcet sounds to listening ears, alas.

Two human sirens, fishers, join our camp
to hear what spirits haunt the rocky cliffs.
Their struggle raises tent and roasts some trout—
hails sweet marmot song with unsteady shouts.
They are quiet when morning marmot whistles
and white-crowned sparrows wake us from sleep
made fitful by the thinness of the air.

Awake, we rise and push on to Boulder—
Byron, Hal and Hamlet wait for us there
and we celebrate each night with Shakespeare,
glorify days with Rocky Mountain birds.

The flatland fields north of city cement
offer us elegant western grebes on ponds
and Lewis' woodpeckers in cottonwoods.
The evening shows scenes of war and peace
in open air where swifts swarm before dark
descends its curtain to engulf our minds
with *Love's Labor's Lost*: I create an arc
that links the pain in my side to the blind
web of necessity that wrecks acted feast.

The oxygen at fourteen thousand feet
on Mt. Evans, rarer still, can impair

the lungs and brain. Here too marmot grunts
enhance the meadow flowers, summer fare
for marmotry. They harvest tundra hay
and whistle while they work in rocky lairs.
The blue grouse with her chicks crossing the road
heralds good fortune. Hammond's flycatcher
calls to us on the trail through evergreens
at nine thousand feet, "I'm here, birdwatchers."
Rocky Mountain high—traffic jams blue sky
of Paradise. Motors hum as drivers
look for places to park. The signs order,
"Don't feed the birds," but gawking tourists thrive
on giving handouts to Clark's nutcrackers.
They cast their bread into the air, up high,
to watch the gray-black birds do barrel rolls.

At Medicine Bow the rosy finches
sit on dirty snow to pose for pictures.
Following faint trails through rocky tundra
we look and find a white-tailed ptarmigan
whose camouflage permits a secret plan
that does not include flushing from her nest
among the gray rocks—except under duress.

We salute the sky, our father,
We caress the earth, our mother.
We acknowledge our cousins
the flowers of the tundra.
Throughout the meadow we search
for our sister, the ptarmigan,
whose gray caress coheres with the rocks,

her softness set in volcanic stone.
Her grey feathers mock
but dress the rocks.

She flies at me with desperate wings
when I stomp upon rocks close to nestlings.

In the Pawnee Grassland we spend the night
in the truck, hearing hoots from owl burrows.
They deride our missing Crow Valley Park.
Too tired to take offense, we take delight
and fall asleep by counting their barks.

Next day we savor some longspurs, McCown's
and Chestnut-sided, and trace these birds down
in plowed fields where they are easily seen.
At a prairie pond dried almost to mud
we stop to gaze at a gathering flock
of shorebirds, where Cedric Foster
and his grandson park their pickup, unlock
the gate and ask us in for closer looks.
We close enough to identify Baird's
Sandpipers among the peeps probing there.

North to the Oregon Trail we follow roads
as straight as those across the coastal plain
though we are five thousand feet above sea
with flat land around us. We turn again
at the interstate and start heading east.

Descending fast from Colorado high,
we journey to join the River Platte

serenaded by raven laughter, wry.
The first-year chase,
belly roll and flail
a strut before a sleeping pal
and a tug on the tail—
the race
is on.
Ravens vault the air
and tumble there
with free falls,
raucous calls
inviting all
to the game.
A test of will,
a means of release,
a complex way
to learn survival skills.
Fond of play,
ravens will perform,
even caged
will entertain themselves
with mime.
Enclosed,
his acrobatic pose
slides down
a smooth perch,
his posture
like an otter's
upright in his lurch—
little boy buoyant
whose mirth totters

solemn cant.
Iron bars
cage performance,
not the dancing heart.

VI Homeward Bound

The distance east we travel fast to meet
the silver arch towering St. Louis
and barely reach the mounds across the river
in time to see tree sparrows well enough
to put them on the list. They are tougher
to find than we expected, but we meet
before dark blears binocular vision.

In Louisville we stop for Shakespeare's
Merry Wives of Windsor, a nightly feat
in the park, where the bard embraces deer
in sylvan surroundings fitting for Falstaff
wearing woodland horns that make him fear
his band of malcontents, who bring us laughter
—wring our minds back to mundane careers.
I long to ask Falstaff what provoked him
to his mid-life folly. "Why play the fool?

Why do they call it the mid-life crisis?
It happens to people of all times
and not at some predetermined stasis."
The question does not identify a crime
but all the same implies a falling off,

at least a mildly inoffensive cough
They all fall from grace though they avoid grime.

Who is they? The holders of what? That jizz
known as community standards? It is
their call. They term it infidelity
and go on talk shows for celebrity.

Careless illicit loves succumb to guns
or tongues. What sins cannot neighbors detect?
Running around on one's mate is not done
boldly, with an honest flouting of rules.
The righteous require a decent respect
for propriety. Those who ignore rules
must be cleverly subversive, not fools.

The pain in my side gnaws against my mind,
a warning birds and bard can only wind
temporary spells. Their magic may not last
to let me subdue the patriarchal past
unless imagination allows me
to fasten mime to timeless imagery.
Tomorrow we make our circle just
at home now that we've slaked our wanderlust.

With and Without Love

Malt, Milton, and Mary Jane

It's on the house,
the keg carried from room to room
is a broom
to sweep away thoughts
that make us believe our life is real.
The warmth of flesh is a reality
we find by putting one finger against another,
the mesh of touching actuality.

Sleek hips and delicate white arms,
five-pronged silver forks
wet with the beer they serve,
a freshly painted carriage
covered tightly with a plastic wrap,
impressed elasticity of patterns
etched in symbolic tracery of clock
with lines of black and red.

Underneath, the whiteness of pillows
and ships sailing
and under this the electric aliveness
of animal azure eyes
in token of the sea
and the star-nippled sky;
The whirling shadows rest
on inundated skin.
a ring to be worn
sapphire and golden in sensuous glow.

A sheet woven bright
for a dance of no finality
lies on her smile;
a vinyl curtain quickly drawn,
the temporary to die
serves the drinks.

Morning Song

Nestled in weary arms
she warms my chest
with youth and laughter
against the coming dawn,
when emptiness
embraces her.

Counting days and nights
before her lips
laugh in my eyes
like holiday lights,

she numbers memories
with sighs.

Recalling how we part
she begs me, "Stay,
oh, do not rise;
the day breaks not, it is my heart"-
light shining from her eyes
overpowers art.

The anguish of the heart
is not confined
to her face alone;
my joy must die apart
from her sweet tones
of love by loss refined.

Night Winds

The African night
settles over acacia trees
leaving me no light
to erase memories.

When the day birds still,
passions no longer quell their cries.
They swell up until
my flesh escapes mind's hide.

Then fever trees
outline on a dark sky designs

whipped by night breeze
for lovers and lions.

Another Opinion

This churning in the gut
is not what
lovers write about
when they tout
bright love in song.

The loneliness, the ache—
the fears that break
out in cold sweats—
I think you will forget
our song of love.

Then memories calm.
They are a balm
to heal doubt's wounds:
cherish quiet hours, croons
my troubled mind.

Chiropractor's Dream

I diagnosed my moon girl
a harlequin some days before
she complained of a backache.
She fell in a swoon
as she sat in my waiting room.

That beauty moon
slipped her disk.
I tried to relieve her with runes
rubbed against her back
while I chanted
ancient spells as I tuned
her body's vertebrae rack.

She didn't like the treatment,
and she flipped.

When she ran, I chased her
over the fields until she slipped
and did not stir as I pressed her hips
against the fur
of our mother's skin.

I massaged and lipped
my moon girl's myrrh
as she lay under the stars where
she had tripped
upon her soft and hidden cares
until the sun awoke
and stripped
the shadows of our moon world.

Bloodline

On telephone wire above clean water
a kingfisher watches for his prey—
for fish in creek, and us, no quarter.

September heat envelops landscape
whose virgin old growth was made conquest
to render the fat of Boone's land, a rape.

Ridge and valley fell to ax wielded
by pioneer—nature could not deny
this great urge that stirred her rocky fields.

This place where we attempt escape
of ancient design lies in sunshine
that ripens patriarchal landscape.

September sun heats my gaze. It gleams
into ageless eyes—kingfisher plunges,
wings folded, beak straight, into clear stream.

Horizons beckon—movement in dreams—
we drain from heights to mountain valley.
Our high desire runs down with the stream.

Silver Linings

We sit and talk,
students who catch flies
or God
upon the pen's point,
impalers and impaled,
we balk
at selling the old lies
with which we're shod
to beg the point.

Old Socrates,
you cry against
our sophistry and poetry
and fees:
your students followed free,
Plato's disease.
With pen and chalk
or hemlock talk
we rap,
we breeze
to grease the laugh
of Aristophanes.
We sense
the mystery of winds,
Aeolian harps,
high recompense
for sin.

Retreat

They talked,
the two from prison
returning home.

One man
told how he'd been
a long time away

From the taste
of woman
and white liquor.

The other man
said hell would crack
before he went back

To the arms
of lawmen
who'd see you sweat, or hanged.

The bus carried them
over the black asphalt
of Camp Thirteen.

Foxtrot

necks bend
as backs sway
got corn
to hoe today
breezes blow
big clouds by
some low
some up high
warm earth
meets the sky
what a shame
we dance to die.

The Catch

Sunspots flickered leaves
as the chameleon
glittered in my garden
and slowly blinked his lizard eyes
before two grimy hands
encircled leather.
My pride of parenthood
was mixed with awe to see
the anole turn from green
to red to brown within
the lizard-catcher's hands—
my son's delight in turning suns.
How swiftly anoles run
among the summer leaves—
those hands that deftly hemmed
the lizard have caught age,
but I am still amazed to think of anoles
turning green to beige and green again.

Yamaha Use

Vrooommm ...
Over the yellow buttercups
I walk outside
to meet my son
announcing with regard
how I, in my name for insurance,
own a motorcycle
I must ride

at least five percent
of the time.
So I am bent
unwilling mime's
affinity to Hell's Angels
and must tacitly approve
an incarnation
of Evil Kneevil ringing his bell
as he hits the groove
in my back yard
churning the buttercups,
Vrooommm ...

Creon, I Have Met You

In modern Thebes old Creon walks
to see we leave the dead in fear;
he walks with lengthened face and balks
at seeing souls cross Charon's pier.

The ancient man still mocks the gods
(or has he clear forgotten them?)
who still demand the turning sods
for eyes that weaken and grow dim.

He has forgot the fate of man,
of Jocasta and his brother;
he lives in Thebes as he began,
rejecting God and sepulcher.

Needed: Pharoah's Daughter

A happy infant should glow
healthy and full of joy—
strong in the life force that grows
from digested lactose. Boy

or girl, love does not question
the future role of diapers,
genetic replication.
Love presses its warmth to fur.

The baby in a hovel
burgeons happily at breast
that's quickly near, picks up well
soon after the squalling test.

Suburbs create marasmus
neglectful as tenements.
The careless bottle can crush—
cold, affluent negatement.

Ackerman's Laugh

Ackerman, the sharp young archeologist,
stood in a cave in Borneo and looked
about him at the walls and read the list
of Vedic heroes graven there by crooked,
gnarled hands that believed in legend.

You gods, he thought, could not keep your friend
alive; he 's been dead these twelve hundred years.
Ackerman almost doubled up at this. He grinned
up at the idols. His face burst into tears
at the thought of this ironic, ageless joke.

He laughed and laughed until he seemed to choke,
until the silent cave began to grumble.
As Ackerman stood there in his witty glow
the ground beneath his feet began to rumble
and idols tumbled down on merry foe.

911: Final Call

Her dark hair in curlers
and delicate fingertips
amazed the men who found her
sleeping with blue lips
near her baby's bassinet
arrayed in pink and white
to match the bright flowers
on the table she had set
for her guests that night—
she had arranged for hours.
Her dried curls depressed
as they pulled her fresh
from the kitchen range,
her unaccustomed stiffness
marked her change

toward formality.
Surrounded by her last valets
she cried less strenuously
about the empty days
of matrimony.

The Door-to-Door Show

Reel I: Sampling

Home from school, looking for a job to heal
the wings of a young bird fallen in flight,
crashed out of college life onto the streets
of harsher reality—winging retreat,
an owl returns with his sorrow toned
to the shade of the arching trees of home,
desiring a Bo-Tree in Virginia ...
contemplation beneath clouds of regret—

with clear eyes Guy sees the need to get a stake:
the want ad reads, "We want a clean-cut man,
personable, young, seeking a career
in advertising. Salary plus expense
account. Apply Montisello Hotel
in room two at nine o'clock on Monday."

No longer there, a Greek revival building
torn down with difficulty to raise up
bare-sided walls, straight lines of modernity
cast against pigeons and valued beauty—

it is there on application day
in June of '53: Guy's plodding ways
begin at eight on Tuesday, on the pavement,
walking the streets, following the chalk
the leader marks to organize Guy's walk.

He sees American birds answer the door
or fly away to hide a body bathed
in startled agitation, like the lady
Aphrodite of the ironing board
who gives him, his youthful eyes, a vision
of American beauty in the flesh, nude,
raising his temperature and humidity
already sweltering in the Norfolk heat.

She appears from the half-light of the screen,
standing at her steaming board, her back to him,
hip curved to the thrusting line of the board,
Venus Callipygous sways to her work.
Casually planted pelvis swings with the work
until the doorbell rings and the nude turns
to greet him, expectant, lover awaited,
aglow with the blondeness of her pubic hair
and the mane flowing blonde to her elbows.

Her breasts lift toward him, the nipples greeting
him before she snorts and shrieks her dismay
uncovered to Actaeon eyes burning
her body. Aphrodite flies away
from glance of boy who offers coupons
instead of the love she expected.

Consider how Tom impaled his Godiva.
What if Eve had flown from the serpent
whose gaze must have surprised her with its glint?

Reel II: Pounding Pavement

From his wagon the leader enforces
the soap. Sample deliveries, coupons
to whet the needs of America's wives:

"We are the boys from P & G
who trek our souls through city streets.
Housewives, hold out your hands and take our treats.
Don't fly away from us like scared pigeons."
Despair, decay, beautify a summer day.

Guy passes by blocks of rubble renewal—
knock at a ghetto door gets sultry voice:
"Coupons for you lady; here you are"—
"Come on in white boy, come into my room
and let me show you a good time."—"No, thanks,
but here's coupons with our compliments."

Ebony skin invites, water steams
in stagnant pools, odors of stale urine
and bacon grease fried earlier that day
mix with the spoor of dogs in the mire.
Sex and nostrils assailed, Guy notices pools
covered with oil slick, shining in sunshine,
all the rainbow colors broken apart.

Two blocks over on Boush Street he knocks
on the door. Short and brown, a warbler
in disarray waves Guy into her room
with ill grace, a deadpan comedienne.
She is confounded when he offers
soap coupons instead of sex—she smiles
and smiles into mirrors without a back ...
a bed, some pictures on dimly lighted walls...
she puts her hand upon his arm and sings,
"Coupons? Boy, now that's just what I need."
Well, Guy agrees, of course, and P & G
triumphs ... a gentle Christ beams down on things.
His youthful amazement complements
her aloe gaze, her saffron smile, her scent.

The dirty tenements, row upon row,
invite Guy's feet to starlings and sparrows
that glare at him from houses peeling paint
as they await the crews to tear them down
for urban renewal. He climbs the steps.

Weathered boards betray yellow paint, tearing,
chipping off, unable to wait for the chain

and ball of destruction smashing its way
along the street toward the water's rim.
The gulls wheeling overhead cry his pain.

He stops upon the sagging porch and knocks
at a gaping doorway, then enters the hall,
hesitant, moves slowly past more doors,
all of them open to his eyes ... feet forward ...
slowly, slowly ... Guy falters to vision.

So peacefully she lies there spread-eagled
on the rumpled bed covered with her body
and one sheet, Aphrodite asleep,
curves akimbo, her silken hair, reddish blonde,
curling under her arms and at her crotch—
she doesn't use peroxide for her beauty ...
a beatific face betrays no sign
of weariness, no lines, no petulance
in the open lips that frame white teeth ...
even in their slight smile they deceive:
wet beauty adorning an ugly crib.

The mounted Venus grins at men as though,
lips parted, slightly moist, a passive dove
awaits the ever-expected kiss.

Reel III: Westward Ho!

At appointed minute the caravan
lurches into dawn, moving south and west,

snaking out on the highway to Statesville
to join other wagons heading west.

Night in a cheap hotel gives Guy time
to study color plates of birds, paeans
to beauty that he fears won't sing or soar.

In two days flight Guy reaches plains of grass
and scissor-tailed flycatchers sitting fence
beside golden dickcissels throating song.

The too-drab plates in Guy's old Peterson
do not describe the beauty that he sees.
The landscape they travel into Texas
reminds Guy now of that Childe Roland rode
toward the Dark Tower. The cattle skulls
offset a lazuli bunting's song
and a ladder-backed peckerwood flying
in front of the car to shouts of surprise—

in the Big D guys lose virginity:
the elevator man pimps, and Lew haggles
him down for Guy. Later, after enough booze,
Guy sees the girl come in the room, undress
and climb upon his bed. Her expertise
staves off disaster, and she stuffs him in
at the black hair that does not match her head.

Long after, Guy wonders what the pleasure was.

Across the border from El Paso,
down the dusty street—the wide, wide street
of yellow dust that leads to Mexico,
a taxicab rolls by the men. They shout,
"Stop!" The taxi takes the jovial crew
to a Mexican caravansary,
a house of rest from the cares of the world.
Pedro swears this whorehouse the best around.
The rates are lower in the middle day;
the American soldiers come at night.

In the corner of a huge room girls sit
knitting, crocheting bedspreads and tablecloths
of intricate designs that remind Guy
of the delicate forms Aunt Sally knit.

"What are they doing here?"—Pedro tells Guy,
"These girls, they make their dowry, then go home
to marry. A girl gotta have linen
and blankets ... these girls work hard to marry."
Red tequila hot in his queasy gut
and this domestic scene whirl in his head:
five madonnas without child sit by the wall.
Then the youngest drops her work to blandish
him. Youth to youth, she seeks his innocence,
but Guy has not found his liquor courage yet
and lets this sparrow from the desert sing
to the next customer, who buys her wings.

Three ladies put their hands upon his thigh,
tequila glows, their beauty grows, and Guy.

Alluring, accomplished, voluptuous,
a lady leads Guy to a room and tells him
to undress ... Guy watches from corner eye ...
as she reveals delightful buttocks, breasts,
and navel above a black brush of wires
that shine invitingly between her thighs.
Her lips are ruby ... it's not lipstick ...
it doesn't come off when she begins
to assault him ... below the belt ... she works
with professional grace ... now Dallas fades ...

O lips and breasts attacking, what were you paid
today ... testing ... Guy finds the past alive
and demands a more conventional love
to save a memory that nothing gave.

Reel IV: No Angels There

Next night the crew stops at a place where dirt
and sawdust fight for mastery of the floor,
the sheets, and every piece of furniture.
But a dollar a night sweeps linen clean
and they do not care for tables and chairs.

Neon signs complement the decor
for the traveler ... heigh ho the wind
without rain blows tumbleweeds up the street
by Jose's Bar and Grill, by a U-Haul It
to the porno thriller on the corner.
Up, up, and over the stucco buildings,
a flock of loud-mouthed white-necked ravens fills

the evening sky. Their color and their mirth
echo Guy's thoughts—fear gives them top billing ...

Here Guy is, strolling down the boulevard
in Hollywood, the glow of Sunset Strip.
Nearby Kerouac, Snyder, Ginsberg, the Beats
are hovering there about the City Lights.
They do not exist for Guy, those poets,
though he has heard the Word of Ferlinghetti.
Their howls mean little to him now—
he knows Joe Page and Joe Dimaggio
and roots for their rivals every day.
The motley birds that move along the street
amaze him with their half-dressed variety.
Gauche, garish, flitting in the sun or lights
from neon signs, these moths and butterflies
must flutter, flail, flame up for poets' eyes.

Right around the corner from Grauman's
Chinese Theater the crew sets up house
in an efficiency apartment, up high.
Danny gets one sofa, Guy gets the other.
Steve and Lew share the comfort of a bed.
Karl looks for quarters to express himself
next to a quiet bar for afternoon drinks.
That's where he meets Lee come from dancing school
who leaps the tables with her lithesome pals
until six o'clock, when she dons a skirt
and serves refreshments to her able friends
in the Rooster Lounge at the happy hour.

The birds of the bar sing early and late,
they preen, coo, dance, and after hours, mate.

Odd birds sample ... Evan bums at Malibu,
spends his weekends with his wife on cheap drunks.
They go down to the beach, light a fire,
eat, drink, until they've numbed their pain ...
then, if they're able, they go to sex
until they've numbed their pain again.
He doesn't pass out toothpaste very fast
and his first week on the pavement is his last ...

George stays longer. He is a future star,
affected ass hiding behind the glass
of his shades he adjusts a bit ajar.
George regales Guy with tales of Hollywood
and has him primed for Jerry Lewis
when he bounces across the street that day
as they set out to sample the suburbs.

The Hollywood Hills solace Guy's Sundays.
Hiking up the street for several blocks
he reaches brush of the watered ravines
and the arid hills that cover their domes
with leafless scrub. Alone, he sits and stares
across the unlovely horizon-smoke
and odd mixtures of every architecture
imaginable to man. Western Pewees,
Blue-gray Gnatcatchers, Plain Titmice, bluebirds
decorate the hills with colorful wings
that waft him two thousand miles in a glance

of recognition. Days without sample paste
restore his sanity—blooms flit the breeze
slipping away from disappointing bees.

Reel V: San Diego

A roost away from home for sailors,
Alice's offers accommodations
to visitors from many nations.

On weekdays when the fleet's at sea her rooms
rent for two bucks a day. Comfortable beds
would give a good night's sleep except for phones
that ring throughout the night-desperate groans
seek the girls who give them comfort ... "Where's
Cora?"
"How can I get in touch with Terry Bird?"
At two o'clock in the morning dark
Guy finds it difficult to understand
their need. Sailors have his curse, curt, unkind
to this comedy of quails. Down at the piers
he watches shags, the cormorants in line
on piles, fishy sailors, comic fliers.
Guy works houses in San Diego light
to offer toothpaste to ladies of the night.
In their jovial mid-day dishabille,
raucous magpies greet Guy laughing loud
and asking what they can do for him—
this sample will taste good in their mouths,
"We need clean mouths in our line of work,
come on in and spend some time. We've lots of time

now. We'll show you a good time today
with the sweet-smelling breath your sample gives"—

the dry air blows through the palms, and he dreams
that gentle female fingers touch his hair ...
laughter rudely wrecks these fragile daydreams ...
he awakes to houris' obscene gesturing.

Hummingbirds suckle November flowers
flaming red along the rocky paths
that overlook the bay. Anna's hummingbirds
streak about to match the flowers with their throats.

Guy sees them on weekends off, but weekdays
he hikes the streets to pass out toothpaste
and watches flashing wings before his face.
Tri-colored blackbirds whistle from the reeds,
a mountain plover flies across the hills,
a white pelican, stately, soars on past
while Guy, earthbound, carries his sample box
from yard to yard.   Beneath an evergreen
he sees a gigantic sparrow singing:
a golden-crown adorns his plodding day.

Fierce beauties, white-tailed kites surprise Guy—
you don't expect to find them in the suburbs—
a California thrasher chortles, loud,
and scarlet finches flutter in the shrubs.

Like the Pied Piper, Guy marches down the streets
of San Diego followed by little kids

who yell at him for more toothpaste, "It's good,
just like candy, good to eat." They laugh
against the toothpaste world of P & G.

Their toothy smiles of glee brighten his day.
The blue sky's loveliness increases warmth
within the heart.   He wishes for a time
to be back in the joyous days of youth ...
heaven paved with terra cotta toothpaste.
The sun of southern California moves
the mind of an Easterner—Guy misses trees
but revels in the warmth he walks for fee.
Ash-throated Flycatchers yell raspingly ...
they, like the phoebes, express a harsh delight.
Steady glow, the autumn sun exposes
negatives inside Guy's throbbing skull
where reel follows reel follows reel ... rewind ...
and the motion picture show grinds on.

He attempts a flight with warblers heading south,
but momentary slips from the movie
show him that this strategy won't carry
him back to old Virginny halfway
young enough to forget what happened.

VI: Heading East

After a long climb the crew rides the peaks
of the Sierra Mountains. Winter sun
shines on the jagged rocks ... blue, red, black hues
thrust in stark lines but where brush shields the sun.

n this brilliance Guy sees Townsend's solitaire
flying-an elegant jewel flare.

On the way east they work a week in Tucson,
a week Guy's mind careens across-crazily.
He rooms with Russ and Steve ... they see him slip ...
he hikes each day and hands out lots of paste.
Guy pushes toothpaste as if it's going out,
not in. Damn, P & G will hate to lose him.
Guy hikes to the desert for weekend excitement,
separation from the paste. The desert soothes
his heaving breast with silky flycatchers—
Phainopeplas—Inca Doves, and claws
of a goshawk set in a rabbit
to end his flight. Aiheeee! Brothers! He shouts
as he senses kinship to them both—
blue sky, blue sky, finality calms him.

Ferruginous hawks sit on saguaro
trees beside Route 66 and hover
in the sky above. Their ruddy fierceness
sends thrills along Guy's neck ... he longs to soar
with them, but still he runs too fast in thought
to feel easy with their circling symmetry.
The Aplamado Falcon pleases him
more because the swift flight on steadfast line
promises an escape from repetition.
Swimming eagerly in the blue he pours
his strength into attack's severity—
rare, taloned bird, Guy wishes flight with you

to shake the grimy dust from off his mind—
but there, you fly, and he is left behind.

As they move into Colorado Rockies
the forests rise about them and the snow
lies thick upon the ground. Forced to stop
for traffic stuck in mush, they clamber out.

Gray-headed juncos fly among the trees,
russet backs sparkle against green and white
to brighten landscape ... solitaires and snowbirds
make irregular yet certain flight to art.

Guy's mind glows with the scenery he sees
at day and night ... kaleidoscopes press him
on though pain inhibits thoughtfulness.

O for a life of sensation, surreal
color, rather than cold sweats in the dark.
If only he could frame his mind with birds
to soar beyond the summer scenes that rush
upon him ... if wings or art can conquer
he may escape the woven plot he fears—

He remembers the lark bunting black and white
beside his small brown mate flitting the sage—
orange, blue, against the purple mountain range
in morning sun ... the sky is eggshell blue ...
the sun warms the artisan, unpolluted.

The picture show across Guy's mind rolls on,
reel after reel flashes frames Guy cannot stop.
As the train sways on through coastal Delaware
shadows from yellow lights and high, dark spires
cast Dantesque shapes over Chesapeake Bay,
match chimaeras within his whirling brain
that lurch over the tracks of sanity—
he can never reach home again with ease.

Norfolk is a few hundred miles away
by railroad car, but he runs in the West.

His late show offers him no intermission.

In the dim lights around the wharves gulls hover

and cry ... if he had their wings he'd fly ...
in the end, his movie is too real for Guy.

Historical Places and Perspectives

Urbania

This whore of a modern world
has high, firm nipples of concrete blocks
and brittle hair of steel that's curled
with the rhythmic sway of hammer shocks.
Under us she lies in her false heat
And gives us a half-hearted kiss
whose acrid taste exudes deceit
and mocks our urbane synthesis.
We moderns love this skilled professional
and buy her on seductive city streets;
her SUV's, stock quotes, and other pals
we crave as addicts crave their killing treats.
Yet we fear her price is growing high
and sense that something's badly gone awry.

Subway Rites

People crowd the underground to move
under city streets an unnatural way.
Mole-like they travel tunnels in a groove
at speeds that might a grubbing mole dismay.
This action does not make an easy play
for genial spirits or for humane talk.
Every day their skies are gray and riders sigh
and many wish that they could sometimes walk
instead of riding subways passing by.
Elbow into rib, foot on other's foot
they beg a pardon and justify flat-
tened arches as the price they pay to root
out the hatred as they hide the impact
in friendly gestures, acts to stay disputes.

Mound Builders

Here, where too many of my friends
have stopped to leave their gifts of waste
I thrust among their colorful bins
to find odd artifacts to suit my taste.

I watch as fragile leaves blanket the rubbish—
technology to which we have entrusted
our lives to create ease and instant bliss
forged in garish forms inviting dust.

Plastic bottles gain brief permanence,
a value littering our design.

Their age betrays our best sense
and rusts the edges of our minds.

Less brittle than a falling leaf
these modern forms smart men have made
of plastic, glass, and metal. I feel grief'
at their longevity, for which we've paid.

Black Mountain: 1930

A man stood caged in his hell
And watched the angry files go by.
Jesse watched the walls of his cell
And heard the buzz of bloating flies.

Some said Jesse boy meant rape.
He didn't have an eye for money.
"Oh yeah, old Jesse can't escape
now. We've paid his lawyer's fee."

Late one night men visited jail
And dragged Jesse out to their cars.
Up at the Gap Jesse looked pale
When he dangled under the stars.

Georgia Southern in Norfolk, 1950

On the street, the lights blink GAIETY!
Inside the front, bare on a stage
Bump meets grind in a grin

For ecstatic animals she's caged.
The burlesque dancer warms her men
And other admiring friends.
She calls to them, quivering, splits,
Swings her mass and twitches a hip
Bringing hands deftly past her crotch
To make a smacking sound of meat hit
Stripped on svelte loins in motion, hot.
Pounding drums and wriggling pelvis
Fire eyes with desire for a kiss
When contortion lifts the light gauze
Mocking her viewers' distant paws
As they imagine what they've missed..
Down on one knee, slim leg stretched round,
Ogling gaze with movements for thinkers,
Sighs, gives together breast pounds
And offers body-stretch glimmers
Of the parts she has danced around.
Outside, the street lights blink GAIETY!

Recessional

When a dark night comes, shadows disappear
into their element except where lights
echoing sunshine bring out chimeras
for those who walk streets under city lights.
In the morning, sun draws forth true shadows
on the streets to stir up candy rainbow
wrappers. Back and forth aimlessly they float
over the skulls lying on the gutter sheets,
over the hands outstretched for a groat,

over the dung of the dogs that stroll for treats.
Rapid walkers hurry by day's retinue
as if blind to a complicity they rue.

Jamestown, 1958

Mosquitoes still sting
In summer places
On the same surface
Of malarial water
Giving no quarter
Producing larvae
That swelled the fevers
In the gentlemen
Who gazed upon
The earlier suns
Of earth's only paradise
Whose later bites give stings
That only worry
People in a hurry.
Now a curio store
Sits on the land
Where many starved
While sitting on their backsides
With fevered pride
Instead of turning a hand
To the hoeing of corn.

Opposable Thumbs

As grail or phoenix feather whirls in air
Promethean hero grasps it there,
audacity for immortality
rends upper air to prove man free.

Fired beyond his slime toward the sun,
he counts stanchions blessings run
through implemented mammal hairs
grace astronaut with palms at prayer.

Caesar's Head, S. C., 1960

American primitive,
old Indian,
he stands out watching
swallows chasing flies
in roller-coaster snatches,
summer wings
obsidian in skies
where leathery things
had flown eons
before totems touched the sky
to signify man's bond
with earthy core.
Still swallows fly
about his face
as new savages defy
wonder drenched in red
to grace
a wind-hewn Phidian.

Dragstrip

Concrete straight-aways on our highways
challenge us to rapid races in the sun,
not only why and where, but how we run
while rubber tires squeal loudly as we play.
Even though we build mighty raceways
for soul-enticing motor motions
with faster speeds to satisfy fond notions,
the screech of tires that tests attempts at tracing mileage markers along our days
with nuts and bolts and rods wears us away.
Racing mind and body all life long
can cause desire to brake and restart
to blend with a daft mechanic's song
to justify the racing human heart.

At Arlington, 1965

Over these three tombs that mark the dead
The marble columns rise up clean and high
Above the trees that tower the Potomac's
Water blue and placid but for boats
Careening here and there in heat,
Their wakes dividing blue with white
To match the ivory columns.
        Quiet has come to their world.

In shade they lie no longer shy of death
Who took the hemlock bomb and mortar shell
To die so fast and easily. In death

They offer no regrets. The three are mute
But do not dwell in silence as the hosts,
Polite and careful, to throngs of visitors,
Multitudes who murmur at the dead.
        Quiet has come to their world.

Better that they are not known,
Not knowing, their oblivion sits high
Above the Capitol. Their classic stoicism
Maintained without the limits of a name
Honors corpses. Unknown they are immortal
And disdainful of the cameras
That click the changing of the guard.
        Quiet has come to their world.

Myths for Space

The men who search toward the distant stars
will think always of home in Akron Heights.
To them their flight will seem a search for sights
more right for thrilling men than Roman Mars.

Yet space alone can breed no Greek centaurs
and men who fly on earthy emptiness
will fail to fill the void they seek to bless
with struggles taken Titan to the stars.

Dateline D. C., Prime Time: March 9, 1977

Blood flows on the carpet from wall to wall
up seven stories and on ground level
painting the citadels of men and God
with crimson on green rugs
to order up an ambulance and pall
in the name of Allah as they create hell
riding out of ghetto deserts in squads
on U-Haul It trucks as thugs

from Hamass shout Mohammed's jihad
that sweeps the streets of the Capitol clean
for killers holed up with sad hostages
heads tumbled out dark windows—
jettisoned women and children deemed bad
drowned in basins, drowned in tubs at the scene
of death in the night whose strife engages
the prophet who spurns his foes

and brings Wallace Mohammed to town to talk.
Gurus of guilt say our society breeds
violence with aid from radio, TV,
and newspaper sensations
that egg on the angry hunters who stalk
down prayers despite cries Islam supersedes
revenge, demands forgiveness' embassy
to unite a sick nation.

Honking horns of morning sun celebrate
the unceremonious end of a show
that filled our void with TV vigil

the politically correct way.
ABC, CBS, NBC rate
this occasion with a corporate glow,
"Did a prophet ride today from the kill
at his ghetto hideaway?"

Lowell Seminar, 1978

Enough for the sweet-tooth bear to desecrate-
*   This open book ... my open coffin.*

Without a hint of guilt they break the comb,
tear the other, black claws attack—the hive
of bees unloosed about their heads annoy
but do not halt the rending paws at home
with honey, the sweet and sticky bribe
to ease the stings stuck through their furry hides.
Their nectared tongues, their colder noses toy
with the circled wax the bee has coned.

Old Lowell, you serve table for thirteen bears
with the tougher wax that dark bees make.
Ambivalent, like flowers, you court
the tearing hands of hungry, growling bears.
Your sugar waits the honeyguide who wakes
those bears to smash your coffin for their sport.

Dream Trip

I fell asleep over Egyptian lore

and marveled at strange lands. Spirits rowed
the boat I traveled in among the stars
and drew me past heavenly sand bars
and fields of wheat fire-shod like pharaoh
on a golden barge the ghostly rowed
to fallow fields under white-hot suns.
The rowers lifted and dipped their oars
in shadow-stroking unison
to send my soul to join scores
who sought Osiris on the brighter Nile
where the sky-high earth awaits grain
sowed by ushebtis on miles and miles
of starlit land covered in cane
upturned by oxened plows on a demesne
beyond the Water Lily Lake
where Anubis measures men's mistakes.
My judgment day still waited when I waked.

Triassics

At the beginning of an age
The awakening actors
Crowded onto a bloodstained stage.
Small, bright
Eyes sparkled
In the ancient forest night.
Lemurs and tarsiers raised their hands
Successively, aggressively,
And debated
The fate
Of the world with their opposable thumbs

While huge reptilian bodies
Crashed around them
Futilely.
The shrews spoke, too,
But the vote was called
In favor of the brainy mites
And they,
As did the Vedic Krishna
In a later day
(when he spoke of courage
and duty to Arjuna)
In the times of volcanic terror
And fires
Exhorted their followers
To victory
Upon the steppes
And plains,
In the jungles of Africa and Asia.
I wonder that great Krishna
Had in mind
We are all
Products of fratricide.

Totentanz

Museum skeletons recall my dream
of sauroids swimming that calm stretch of ocean
called Sargasso Sea.

They bowed their necks in question marks
with rhythmic grace as they swam
in unison to land.

Swimming to the beach
they crawled awhile away from ocean
before they left their bones in muddy streams:

Those swimmers now are dead
though we walk on erected
to note their drowning on the land we tread.

Primal Question

Was there ever foolish man named Adam,
reluctant wearer of a fig-leaf cover?
Did he bug out of Eden on the lam
or was he myth of some burnt Hebrew lover?

Modern Technology

Pleistocene bears and lions roamed Europe
and disputed the rule of the land
with mammoths and Mousterian man
who huddled in the bear's old cave and cupped
his ears to hear the roars echoed the hills.
He dreamed of daylight and the coming kill.
In the light of day growls frighten less
as he fingers the bear's teeth that will bless
the plan of a stone-and-muscle engineer

whose technology will slay the giant bear
or marauding lion whose mane he prizes.
The fears he feels he hides for he despises
men who throw their weapons down and run.
The rule of technical man had just begun.

Embarkation

Jellyfish and trilobites flowed free
in water womb while Icthyostega bore
his bulk to bleak and cold expanses
whose psilophytes urged roaming more.

Out from watery edge, therapsid tree
turned double eyes on *terra firma* shore
whose moon and star and planet dances
beckoned them to roam some ages more.

Up from Devonian coasts, we stand free
with mind on past and future shore.
Inherited primate curiosity
prompts us to lift ourselves with rocket's roar.

Paleolithic Ax

The form inheres in the stone
where the poet seeks its shape
by flakes with which he can hone
from rock the poetry of blade.

Children of Daedalus

Hut gave way to palace
and pyramidal grace
and the donjon keep
to skyscrapers that leap
to clouds in a concrete race.

Carefully worked stones
followed implementary bones
that gave way to swords
forged for tyranincal lords.
We have not lost our urge to hone.

Old Architects

The ziggurat towered bricks
of Sumer, Chaldea, and Babylonia
betray the waste to our newer tricks
although we're somewhat stonier.

Because he dared the priests of Bel-Marduk
Old Nabonidus lost his empire
to the conquering Persian fire.
He spent his time digging for record books.

His searches conferred no success
to Belshazzar's army, poorly led
and unprepared, falling defeated
before the swords of Cyrus' conquest.

Despite Nabonidus' time at play
while his empire burned,
his ancestors had learned
to build substantially with clay.

Had he paid heed to the lessons
of those who had created bricks
he might have known to rely on
their clever architectural tricks.

Jilted

Dido mourns and thinks of his caresses
enthralled with his touch even on her pyre
and echoing his shape in leaping fire
as the flames eat flesh he still possesses
while she thinks of Trojan sails in the breeze
and finds no memories that can appease
the tragic mask she wears in the light
of his deceit. She sings her last song tonight.
With a nod to the gods, Aeneas' love
sails off to destiny. Her anguish sings
across the waves to blend with cheers that ring
from the throats of his sailors who shove
their oars into the sea. The woman sings
to the flames of her banquet's fiery love.

Thanet Isle, 449 A.D.

The blow fell, the empire broke asunder;
horsemen and hunters pushed on weakened Rome
luring the leathered legions homeward.
They left a half-tamed land to melted men,
people untaught of spears, unable to sing
the ancient Celtic lays of race-birth.
Then, over the wrath of northern oceans
out of fjords and coves came bold seafarers:
Hengist's men looked at their haggard shores
and longed for the rolling British hills.
In their boredom they recalled chalky cliffs
as mead cups lay drained in mead-numbed hands
and the bard's harp strumming raised no head
to listen as he sang of Scylding's deeds.
They sobered, then gave rise to bloody plans
that set the dragon ships upon the sea.

Anthropology Lecture

"Students," proclaimed the careful pedagogue,
"no doubt you've heard of Arthur Pendragon,
who for many centuries has set the world agog.
For hundreds of years he's raised his flagon
in gilded, unduly romantic stories.
In fact," the professor paused for emphasis,
"the literature is full of him and his ploys
and how his knights heroically defeated lists
of evil men and ogres who deflowered and devoured
British maidens in their lust for power.

In all the lands of fabled lore
there is no king like Arthur, none is spoke of more,
or lied of more convincingly,
but the truth of the matter seems to be
his sword, Excalibur, wore gaudy bands
of polished Roman iron, and the wielder's father
had been a little loathe to see the Roman
legions head for Gaul and the Imperial power.
So, there you see him, a hand-me-down swordsman
with a deep-gorged brow to stand
upon a barreled ton of sensual flesh;

the noble Guinevere was a camp-girl fresh
from the fields who loathed her brutish husband's
drunken ways and daily dallied on the side
with that smoothie, Lancelot. She cried
when she thought of that lump, her husband.
Knowing this, how can we speak of Camelot
as a misty, storied place without a blot
on maidens who took their troubles there.
Arthur met the Saxon, not in stoic calm
But in a liquored passion fired by the balm
of gold his fellow Britons paid for care.
Artoriosus, Bear, was what a querulous monk
called him. Bear struck terror in his foes
because of massive limbs and strength—a hunk
to solve some sluttish British maiden's woes.
He dwells in romance now, not history, the glow
that fabled lore can spin around a hulking brute.
Marvelous tales are told of how he strove
to amaze men with greatness of repute

in a world where evil men were rendered mute.
Ladies and gentlemen, we scholars know
his heroic struggles were self-centered,
not the noble acts admiring poets entered
in tales of Lancelot in Celtic lore."

Dead Scholarship

On travel posters from the past
and in the newsreels now and then
tall, vine-covered Angkhor Vat
threw shadows with a solemn cast,
spread broken temples in the sun
and threw down pagoda patterns
shining from reflective pools
lying placidly and quietly dun
beside hoof-marks in trampled burns
where the shouting peasants
and their water buffaloes' feet
have churned Cambodian jungle ferns—
at least until a red star rose
insisting scholars march their feet
to bony fields of red deceit
to supersede the buffaloes.

Yeats' Statues: a Commentary

Pythagoras gave the Grecians factors
To make the hair, the lips, the eyes to move
By numbers wrought in marble character,

Lost on all but boys and girls in love
Who dreamed a fleshly statue's grace could act
to hold the love they feared an artifact.
They pressed at midnight in a public place
Live lips upon a plummet-measured face.

So they grew to sensuous godhead, the youths
That believed a chiselled stone or frieze
That looked like actual flesh could stem the truth
Of oars upon the Hellespont, that these
Emboldened numbers wrapped in rock could drown
The million oars of Salamis with stone,
That imagined sense could defeat harsh sounds
With ideals made manifest in sculpted bone.

Image of flesh wounded abstraction's cage
But marched to an Indic tree where it grew fat
And lost humanity. Victim of hydra rage,
Conqueror was conquered. Icon sat
Inhuman and gazed with no urge to know
The wisdom of unreal and pulsing number,
Avoided sight of eyeballs that enjoy a show:
Empty eyes found meaning in self-slumber.

When Yeats summoned Cuichulain to his aid
He did not call him from the eastern land
Where Gautama sat blank and unamazed.
Yeats called him forth from olive groves at hand,
From Irish Greece where modern heroes stride
Upon the ground that Celtic blood has dyed.
Yeats searched for symbols there that he might trace
Live lips upon a plummet-measured face.

Intelligent Designer

From out of a whirlwind thundered epithets:
"Do not blame God for pious blasphemies
uttered by proud fellows, My failed prophets.
God tolerates messengers such as these
who think He cannot work divine mysteries
that evolve as men's comprehension lets
them realize their Lord's complexities.
My words are worked awry by false prophets.
It is not easy for proud men to see
as cousins those who grin like chimpanzees
or those who create bowers like fishnets.
I'm the nine hundred pound gorilla, right?"
Trembling, I answered despite my fright,
"I believe in God the Father Almighty
and in Christ, His only begotten Son."
The thunderous voice replied, "It is your right
to doubt those who preach rigid religion
proclaiming nature's God implicitly
gives them the wisdom to eliminate
the changing concepts of evolution
and circumscribe mysterious origins
by denying ancient life while preaching hate.
Do you rule God? Do they? What rash men
puffed up with pride set limits to their maker?
Did they create a universal order?
It that first thought about God became man,
Why carp on what flesh came before Adam?"

Surreal Songs

The Maker

The poet must dip into his soul
(or what Freud terms the Id) to dredge out
mysteries that lie in wait for their parole
from jail of nerves whose crooked circuits flout
the self: to look, ask, force words to resemble
emotions that show red impress
of human flesh in both silk and burlap dress,
emotion taut and not dissembled.

Humanity revealed is half the fight
for freedom of the thoughts he truly feels
but must divide into artistic light
and depth and form: his consciousness congeals
a clot of words to stop the flow of thought
from passing back into the shadow world—caught
by making word curl about idea to wheel
forever symbol, forever wrought.

Kinetics

No marabou for us
who do not care to understand
why the devil fell to dust
or that we sit on earthy land
halfway between chaos and hell
where artifacts made to sell
satisfy no man who will not curse
a providential hearse.
Still, we count to three
to seek divinity,
but find perfected actions
in chain reactions,
at blinking bar ways
and neon copulations,
our sins revealed by scientific play
in a tube of DNA.

Hedonist at the Madhouse

Pour qui sont ces serpents
Qui sifflent sur vos tetes?
     Andromache (mad)

Les liocephalles, pal, lizards, not snakes.
Crawl down to the pit (quit), live with your friends
from San Francisco, the brave young poets
who strive (thrive) just to stay alive on dope.

They see hipsters.

Hipsters witched my brother. See that wiggle?
That dance. A trance. That's what they put my friend
(covered in soot) into. Who? The young man
who (to rue) chinned himself on the tavern bar
(his star that guided him after lights out).

Poets have the favors of that doll there.
What's her name? What *is* that dame's name? Circe?

You bet, brother, no other broad's that game.
Circe, yes, the temptress, like no other
Girl who ever owned an island (don't pout
About it man, she'll do another dance.
Her real name's Helen, see that abortion scar?)
She makes those poets sing (fling out) their song:

> *We live on jimson weed,*
> *It cures our insatiable need,*
> *My brother bore a stalk about;*
> *You don't need to rant, just shout*
> *Your tale told by your sweet mother.*
> *We are not idiots, we're just hooked, brother.*

That *was* a batch of bull, here, have a drink.
Their tale is full of heartache. On life they flake.
You know, it was laughter at it (Attic)
That gave Greek gods (those odds) the bellyache.
Maudlin poets ('Terpe's pets) are not tragic.

Killing Time

Our cards are played with little care.
The lights that strike us on the run
from solitaire to soaking up the sun
hint an instrumentalist of air
has missed a cue or muffed a pitch somewhere.
The lights
of red and blue and white
that splay
our roaming holiday
give us fright
of pause
and warn us to find cause
we pirouette in flickering stars
declaring instrumentalist of air
has missed his cue or muffed a pitch somewhere.
To forget our internal wars,
we play another game of solitaire.

49er: De Gustibus ...

He wishes for shade, not shots of sun imbibed
By the tall, sad saguaros resigned
To standing mute on parched and rocky earth.
A lean and naked arm gives little birth
To water under pebbles scraped away
As it digs underneath a river baked to clay.
Rocks sparkle like gold in coffer box.
Oh my darlin' Clementine's gold locks..

Imagined drunkenness of rocks and sand
Mockingly defies a scratching man.
Sensitive fingers sense their act's undone
By rocks sarcastically aimed at the tongue
Some god has swollen to grate upon a cheek
To curse inscraping hand along dry creek.
Rocks rattle underneath a creosote bush.
Oh my darlin' Clementine's warm kiss.

Sails and Indigo ...

lips carry heavy cargo
of bunting and indigo
schemes, in touch of deep blue
on skin's pink cockatoo
for the neap envelopment
of those thoughts present
on sand under the bend of the knee
and the wavy force of the sea.

Flambeau ...

glowing sun-red ships sailing
with cargo to leeward,
to Sunday dresses galing
on slim-waisted girls tarred
by the storm clouds cowling
close over the schooner charred
by my love's hair trailing
in the wind's force, fiery, starred.

Road Race

We are the lads of little deaths
in love with a girl named Thrill
who funds our motors' auto breaths
and races us over the hills.

We think her sporty car a van
to move us fast from morning's sack.
We flee another seductive scan
on her maniacal rack.

We slip into our curves beset
by our girl with mechanic guile
who smiles and flicks a cigarette
as we drive into the night in style.

Liftoff for Mars

Not knowing why, they placed me on the pad
while moving stars proclaimed nativity
beyond their lovers' dreams whose lives lacked space.

After the countdown came, the rockets flamed:
screaming, into the clouds I soared at ease,
unseen the cords that held me to the ground.

Nine months with stay-connected rope
umbilical brought blast to push me forth
with jets programmed for murder myths.

War Games

Armored column speeding hard,
Parsifal thunders to heel
Now dismaying castle guards
In fear of clanking wheels.

Parsifal's howitzers clamor
Against castle walls redoubt.
Their music plays with candor,
No chivalric peel or shout.

Bold, destructive guns subdue,
Shouts heartless steel. Those inside
Quake. Heavy metallic blue
Threatens with a clanging stride.

Castle bridge rolls down to cold
Steel tracks. Parsifal falters
On beleaguered stony threshold,
By Molotov cocktail altered.

It's A Jungle Out There

A rajah omnipotence sways splendid
Indian elephants guiding his intended
Through a monkey jungle alive with howls,
Owl calls, tiger screams, frogmouth growls
That break the sleep of Maharanee's night.
Nightjars and thrush jays duet in moonlight
Emitting liquid sounds that soothe her.

Languidly the princess' musicians stir
Their strings. Their rhythms embody craft
As they sound through monkey jungle miles
Whose animals hunt while deadly arrows whir
A vengeful ire on gleaming yellow shafts.
Under sunlit skies, unheeding guile,
Swimmers stir the jaws of lounging crocodiles.

Trips

The druggings of our inconstancy
lead mixed-in-acid brains
from the squeals of infancy
to the tracks of insane trains.

We desire a colorful ride
while traveling all our world
and gathering wealth besides
before the last train has whirled

its demanding, mocking wheels
to the heated invections
and insidious peals
of destiny's drugged directions.

Man's Buddy

In our feed bin sits a defiant rat
who curses the claws of our barnyard cats,
eats his fill of grain and grows rather fat—

a magician of meanness and rabies spit
with ability to give us shivers lit
in the redness of his eyes, narrow slits
that leer with evil, apocalyptic glow
of small brain cunning that lets us know
here is an animal who has seen the low
side of the universe and that those teeth,
which chew down on corn in pouches, give sheath
of muscles a cutting strength, a wreath
of sharp death that has played with flesh
as well as eaten the fat grain fresh
in the trough or left where the chickens thresh.
Snarling, murderous magic of rat
sits on his haunches and shows us that
he's not afraid of man's god damn cats.

Sharing

When a wandering hero
asks hurt men to sing their woe
they cause his eager ears to ring
unpleasantly no doubt,
for they quite loudly shout
that he must share their stings.

Sans Sense

The Buddha's rhinos walk alone.
Their master has taught them to cut
Emotion's excess to the bone,
Good disciples almost brute.

They seek to reach a senseless state
By thrusting off their sensuous foes.
What sorrows do they see await
Bold men that shun cold placebos?

Useful Catbriar

Uncle Remus termed a briar patch
the downy nest
where ineffectual pests,
droll and clever,
elude deep scratches
with animal endeavor.

Luther wrote, "Es gibt ein Fehler"
and posted his 99 articles
offering impolitic particles
of truth for paler
souls who didn't know how high
a devil to defy.

Should we appreciate a friend
who pushes
us to a thornier
contemplation of the path's end
by using bramble bushes
around the corner?

Cold Tracks

On tracks of limbo fled my mind
Along the railways of the brain
Where all the stations on the line
Stood dark to my locomotives.

Images of train and track lit
All my dreams. They would invade
My sleep, but they could never fit
In. My mind refused its function.

So I sent engines out to junctions
Far beyond grim railways inspired
By my engineer. His conjunctions
Were too bleak for me. I had him fired.

Lessons Unlearned

Plato's apostles
go importunate lengths
to set our brains ajostle
with eternal drinks
that offer unseen
the bright ideals
always demeaned
because men feel.

Family Tree

Is it a mortal sin
To deny a cousin?
We love mysteries
And learned arguments
About the number three.
Who would deem it evident
Young Eohippus gallops
In the best of beds
And chimpanzees
Easily learn to chase
Slender Jane Goodall
Around a jungle tree
To pick her fleas
While filling their stomachs
With store-bought bananas,
Or, impulsively,
Deign to eat meat
Tearing flesh from the bone,
Smacking wide lips,
Then belching politely
An oriental primate contentment
While sipping their tea?
Now we must say
That ninety per cent
Or more of our DNA
Claims much ancient
Evidence of family descent.

Deer Sighting

From flowered roadside
a startled doe sprang
through the green yard, ran
into the house hide.
The flaming red sign
of her head arched high
seared my tinder mind
with running wildfire.
Road kills are turned in,
not legally kept.
When their fires burn men
they waste heart's unrest.

Idealists ...

Like weeds cut for the sap to run,
our hopes fall down at a thrusting scythe
whetting their crowns from stalks that writhe
while thoughts shout; one or two snickers
mark the ended bouts waged with cankers.

Urban Olympics

"I didn't like the way he rapped and jived,
but he looked Olympian in that dive."
Just like a pal at ease, he grabbed my arm
And caught my eyes and my shrinking alarm.
I thought him one of the bums from the park

Whose bottles crawl with bugs and gutter garp.
"My buddy jumped off that building, up there.
Like a skydiver, he flew a minute
At least. He used no prayer or parachute.
He floated for a moment in the air
without a cord or silk. He was astute. It was a hoot.
Another day I'll try Olympic air.…"
I nudged aside a ragged man's despair,
Fearful of his eagerness to gain repute.

A Final Solution?

In days far past recall of us
who wander down the modern maze
some men of Greece once wandered thus
while Buddha sat in seery daze,
but sages waking now will die
to leave their questions in their stead;
like other searchers long gone by
the Buddha and the Greeks have fled.

We wonder why they sought to cease;
we sigh at them but think we shall
uncover truth. To plead a lease
on life adds zest to our morale,
but Hamlet's qualms at suicide
no longer teach us to abide.

Dental Evolution

Tigers among us bare great saber teeth
dripping glowing blood and growing long-
er as they slice up dovish prey whose grief
increases tigrish appetite gone wrong.

I recall that teeth too long may shrink
From lacking meat. Empty mouths and growling guts
are what caused saber-tooths to grow extinct
when tusks grew longer than the prey they cut.

One-way Trip

Nina, dear, sign up for a trip to the moon.
It's all the rage. We're leaving soon.
What's a million or so of earthly gain?

I wonder. Do you choose your captain?
If he's a hunk, it might be worth the money.

Nina, we mustn't bother how he looks, honey.
Women our age don't complain like a whore.
You must abandon complaint. It's a bore.
Like an old corset, it's too luxurious to be funny.
Times have changed for lack of space. Our times anoint
A society out of joint with our stars. That's a point
Poets have pummeled for centuries
Of fire and ice, but now it's our disease.

Spacemen have nothing on that cow who leaped
So high she circled her neighbor's sphere and reaped
For mankind condensed milk in our coffee cups
And dime store gadgets and satellite pups.

Let's go. In space ships we'll limp with creaking joints
To our final dim, uncharted point.

Old Wives' Tales

Ghosts come at odd hours to haunt men
the old crones used to tell the children;
but ghosts are not always confined
to haunting hours assigned:
we march through shadow days; content
to hide from our dark nights,
we erect electric monuments
to drive away the witching time's suspense
with incandescent lights.
We cringe beneath a hardened shell
of toughened flesh enveloping old fears
of witches tending flames to sear
our shadow world with chants of hell.
We are like locusts grubbing in the earth
who seek to cast off a chitinous shell
yet cannot climb to sun to find new birth;
our hidden fears we cannot admit
to our neighbors' prying glances;
ambiguously blessed, we move our spirits
round and round in an internal dance.

Campfire

Burnt by the sun the savages sit to eat
Around their fiery circle's leaping glee
Upcasting shadows over sand and sea
And canines lurking to devour a bony treat:
The men and women squat to gnaw their meat.
Their raucous laughs and primal grunts declare
This beachfront campfire is an old affair
And glowing embers underscore their feat.
Sage squaw and willing mate will soon deceive
What ancient marriage rites thought to relieve.
With fierce fun lustful glares they pair
To seek their blankets' contraceptive care.
Their hot rites have an ancient heritage
But these savages savor a new age.

Desert Blooms

Cowboy Judas
Rides a radioactive horse
To his wrangling place,
Rides atomic hearse
From the Eden of first curse
To billowing clouds.
Shrouds cradle our sun
In a hydrogen
Annihilation.
Crucifixion and hanging
Served as archetypes.
We learn to ease our angst

With handy wipes.
We are blown away
By small children's games:
They will count before
Coming, ready or not.
Three, two, one, ZERO.
The ground is getting hot.
One, two, three,
Here we come infinity,
Our bid for divinity.

Ostriches

From donning synthetic satin
to the factory clang at matins,
we hide our thoughts to ease the whir
of car and cocktail pother.

Forgetfulness is bargain folly,
an inexpressive dolly
on which we moderns seek to roll
away our past, and soul.

Circe's Lover

Foolish boy, why do you tremble so?
You loved me, you said, and my wine so much.
Stay here and taste my deathless love and grow
Like an immortal god. I hate to chide such
A glorious youth as you, but you must

Share my bed of lilies, my marble halls flushed
With Pompeian pink. I will not let you go.
Stay here with me and revel. Do not go
To her, to Persephone. She's cold
And has a husband too. Let my profession
Of desire, my call to frolic softly fold
Away your inconstancy and make us one
Again. I will not ever let you free
To love me less than cold Persephone.

I love you well enough, dear Circe,
But you can see I need Persephone more.
My thoughts are ever of her who thrills me
With the soft black color of her hair. Her doors
Are always open to her lover. I live no longer
On account of wanting you, sweet Circe,
For I am enthralled with a stronger
Mistress, one who will not let you delay me
With your sighs. I give you up, my mistress,
As did brave Ulysses. I leave your wine
Although I love you as my life. My empress,
You must drink the liquor of another's vine.
I love your sun-kissed arms, dear Circe,
But my drooping hands have found Persephone.

Paternity

Father hocked peace many years ago.
He waved as he marched off to fight.
He knew a price was on his foe
Just like any store-bought item.

He never came back from war to say
Litter bearers carried men from trenches.
The cost turned out to be the solemn day
They bore the hero's corpse and stenches.

The Music Contest: Retrospective

Thousands of years ago I sat down here
And played a song accompanied by thrushes;
Then young girls came to listen, shyly to peer
From banks grown still except for rushes
Swaying to the rhythm of my silver lyre,
Told with music how my worth had been maligned,
Told how I, Phoebus, was moved to ire
Against deaf Midas, that humbug heretic
I gave a prize of donkey ears designed
To show him what he knew of music:
The pipes of Pan sounded pleasing tunes,
But Apollo plays mainly for Muses.
Pan made merry marches on the reeds he'd hewn,
And Midas thought music *only* amuses.

King Midas sat down and judged a while;
At last, after many a thoughtful nod
He said, "Sir, your tunes have a sort of style,
But to me they sound quite strange, so odd
I must disagree with Tmolus and his sneers.
I, Midas, award the prize to Pan."
So Pan got palms and Midas got asses' ears,
The proper things for so dense and dull a man.
He hid them under a special cap,

Swore his barber with dark oaths to enwrap
His ears in secrecy, but desire broke bounds.
By this stream the barber bent and whispered,
"Midas has asses' ears." The rushes heard.
They grow each spring to murmur revealing sounds.

Pilgrimage

Reborn on a psychiatric couch
from the womb of an easy-virtued shark,
an optimistic Freudian sailor slouched
down to sail as helmsman to a barque.

He set out seeking brave new worlds
and dreaming of a tropic paradise
where he could satisfy the ego hurled
too often at the bias of the dice.

The eastern shores receded quickly
from his modern Telemachian ship
stringing lines and foam out thickly
from the tiller in the searcher's grip.

Awakened after somnolent ocean days,
the seeker encountered an untamed heart.
He deemed his journey gone astray
and voiced regrets for a misread chart.

Environmental Hues and Blues

Winter Fare

Wings hover silence as a harrier
slowly seeks a mouse or vole;
to keep his slim form from hunger
some other life must pay the toll.

The gray wings fold and talons drop
Upon the prey whose cries transform
The whitened world in bloody swap
Beyond tears, as form gives way to form.

Passion and the Desert

Full throttle
came the goshawk in his hurtle
at his prey—
ten eons and ten feet away

caught sight of me,
applied wind-scuffing brakes.

He looked across man-million years
of fire and gods
(I thought in fear) with fierce disdain
before he circled free
and shied away from me
in search of rabbit prey.

When I turned back
amid the cross-like cactus
to my coffee by the fire
and my stale taunts of god,
I knew myself, a liar.

Yard Care

Whacking away
the wild things
my neighbor manicures his lawn,
he is retired
and cutting weeds is something to do.
His dandelions are dug up
while mine run wild
with a mass of green and gold
whose yellow hawkeyes
dance in the breeze
on the spring lawn.
His tulips and marigolds bloom
only within well-tended beds,

and any tree or brush
that defies him
by getting out of bounds
invites a sharp ax.
He tends his yard,
not mine—
I desire some days
to yell at him
to let some wild things live.

Peccadilloes

In my dreamland locale
there were times as a child
when I rode the hogs in glee
over the hilltops and the trees.
I rode all wild in dreams
whose colors glimmered like the gleams
at sparkle in the boar hog's
small and flesh-ringed eyes.
I rode red hogs and green
with black and white between;
mixed in color and flecked with mud
the brilliant hogs ran
to the naive emotion
laughing at a pastel notion
that peppermint peccaries demand
implicit statements of desire.

Links

Curved wings hover escarpment rocks.
A fat rock hyrax whistles fear
of their shadow. The whistler's shock
pierces the ear.

The dark volcanic rocks embrace
an eagle's nest. Humans ask why
an elephant's cousin should race
the eagle's sky.

Curiosities of nature
abound in ancient Rift Valley
rocks. Not all eccentric creatures
ape manliness.

A Substitution

We are a picnic people on the run
who stand awhile upon the sandy cliffs
and stare across the ocean far away,
watching clouds form hieroglyphs
around the punctures of the sun.

At our backs tall buildings
poke their towers high and stark above
the furious honks of motor horns
which symphonize the loves
of small and frightened worldlings.

Across the water lie the golden isles
where happy savages danced and sang
about a careless sort of life
until the apostles of steel fangs
substituted cans for coconuts and smiles.

Cumberland Gap

Hiking the forest,
in its half lights
the blinking actors recoil
from the shadow sights
as they toil
with their beasts on the hoof.
They complain
about the miles they march
yet disdain to ease their act
and rest tired feet
in bivouac.
Led by the restless scout
whose tales taunted their beds
and set them west on a head-
ing—stilling their doubt
and fear of losing hair
from something far more savage
than care
or pulling it in dotage.
They follow Boone's blaze,
their stars twinkle, then sprinkle
a geologic condiment
into the maw of stony earth,

that old crone
who spouts her excrement
to build monuments
of rock embracing us all,
even those who drag-
race dreams over the hag
in motorcars.

Grandfather Mt. Salamander Hunt

Grandfather's *Plethodon welleri*
encompasses blood, rebinds the fiber
once served giants of dragonflies
and beetles now encased in amber.
Where spruce and fir and yellow birch
veneer the floor, I seize moldy crud
to study wriggling amphib urch-
in's legs freckled with ancient mud.
With zigzagged streaks of gold and green
Weller's Plethodon casts a spell
amazed the man who found its sheen
to gain its naming for death knell
when Weller's steep descent to death
footnoted salamander breath.

Homage to Ruskin Freer: Naturalist

Rose, indigo, ochre birds on Blue Ridge
peaks, maple stirrings, flowers on mountain fields
paraded in his column's friendly gaze;

imparting them to us he built a bridge
to nature's cornucopia to yield
Freer gifts when he explained its ways.
He took me out one day, spring sun ablaze.
At leafing Otter Peaks we shared afield
a Lincoln's Sparrow all too rarely seen.
In rambling woodlands, he found cause to wield
his pen to take us to the Natural Bridge
to watch kalmia and rhododendron preen
their flowers white and pink and set in green
for those who wander paths along the ridge.

Evening Movement

Traffic whizzes by row upon row of cottages
edging their boards toward the beach.
As shadows creep along the boulevard edge
in late evening appears a lonely girl
to touch the cool wind blowing from the sea
down to the brown knee of the sand.

Stooping, preening a reluctant curl,
she tries to place it as the wind frees
the hair from her clutching hand.
Jet trails etch across the darkening day
as the sea foam ripples about her feet
and an oil tanker lumbers across the bay.

The reach of sand shimmers like chrome
while her footprint in the sand retreats
leaving an undergarment of foam.

She watches the raucous tern on raid
diving for fish in the bay where pilings stand.
It glows roseate in the dropping sun.

The fisher's trade he plies from piles at parade
rising waist-high out of the sand.
The tern drops straight and rises on the run—
wise and selfish for its downy young—
from the ravening gulls; eluding maws at hand
with swift strokes he races home before the light.

This fish will feed the craws of terns tonight
while the girl's foot arches in the touch of twilight
and she savors the salt air galing
as the water reaches upward in the night
to kiss the evening stars now trailing
out their laces as the girl seeks stasis.

Modernity

Over concrete walls dim suns may rise.
All day a busy city wrings its hands
while choking the gas of its demise
in the dank streets where the traffic stands.

Signs of life, the snow geese in their flight
move past the evening fumes in a wedge
that commends larks and jays and woodland hikes
to those gasping under concrete ledges.

Early willows give suburbans hints
of life by signaling in cyclic code
as returning indigo buntings sing
from telephone wires that ring the roads.

What armies of suburban sprawl are marching
to damn the deadly fumes at city halls?
Who heeds the hint from hawk and swallow wing
riding the winds and circling the thermals?

Prologue

The soft whir of wind
postures a bend
and formal bowing from the pines
on March evenings
when the small frogs peep signs
of seasonal yawning.

Farm Lands Remembered

Breezes blow up lightly in the flatlands
to ease the humid calms of summer suns,
throw the red-white tassels on fuming corn
amid a rustling in the heavy fields.
Then dark clouds ride and thunder peals.

We hear voices out in the summer crops,
the kingbird's defiance of the raiding crow,
the quail in thickets talking low together,

the meadowlark's whistled anthem to the storm
over wettened dirt in sodden grain fields.

A red fox feels through wet grass silently;
insects shake their soft-strong pedals, toning.
When the land is laid in, broken, bare,
for the seed to hit the fallow dirt
we pray a good end for its fertility.

The armies of suburban sprawl advance
against the troops of the ditch bank salient
whose battle lines are firm before the steel
machines dig up the boys of gaudy green
whose waving locks live on in memories.

Wetlands in the Mind

We sit beneath the sparkling sign
of Buddy's Pork Barbeque
and lift a can of beer
in honor of the wetlands in our minds ...
over the sunset,
out of the gate,
burning ochre and crowned
with agate
comes the night's pet;
a black cat
leaps through purplish mound
to yellow mat
of smoking clouds ...

Around us blink the evening lights
of neon announcing fun,
movie stars and old age in the sun ...
while the egret,
creamy white,
stretches out dark legs,
circling round.
climbs from dregs
of water
high
to burning sky
and, silver in the light,
bathed and wet,
flies through the smoking clouds,
until the black cat scratches sun
to make a kill
of Eden's glowing horizon ...

Then night birds
hunt for moths whose herds
contest the sky
with neon pyres
lit to service my
Flight four-o-twelve
and beacons of infinite motels.

Voyager ...

Consider, confess a cosmic joke
in the subtileness of southern sea sun
against the corals made for other oceans

in service to quaint dinosaur folk,
not our Dionysus of mushroom smoke
or our sleek supertanker nations.
Consider, uncover the comedy
in universal cannibalism of our sea;
its dainty, embellished albatrosses spar
with biting white species of sharks,
each preying in no-quarter war
on other creatures' hides they mark.
Consider how slowly the waves wear,
tearing coral away with the tides.
Noah's flood has not subsided;
typhoons roil two-thirds world's despair.

Environmental Aesthete

I want to grasp my fancy
by her chemically washed hands
and pull the lumpish woman back
to a mystic vagrancy
beyond the isles and lands
of Mab, to a grass shack.
Birds as well as planes have wings,
but skyscraper man wants props
bought with industrial gold
and learns to despise the stings
of mosquitoes as malaprops
from a life covered with mold.
From my gardens stained with blight
I will try to harvest cankers
coated with a metaphor that's handy

substitute for Miller Lite
to replace the man-made, ranker
weeds, whose bitterness gags me.

October Evening

Above the still leaved trees
a silver eye
floats on oceanic sky
waiting for evening breezes.

In the autumn twilight
random footsteps softly crunch
the radiant, fallen leaves.
Failing sunlight
glows on the lips of thieves
who have stolen their lunch
from the forest's delights.
Birds have traveled beyond
the sparkling breast of an autumn pond.

In an old tavern under the tall trees
a sad guitar offers a song
below the dusty stairs whose creaky pleas
complain its boards have worked too long.
As the night falls a stranger
hears the music floating to his chamber
as sun fades to black velvet under moonlight
enveloping the sounds of night
recalling hints of danger.

Above the still-leaved trees
a silver eye
floats in star-studded sky
waiting for morning breezes.

Seedtime

Red, yellow, turquoise colors catalog
returning swallows, gnats their emblem
imprinted on the mind as man and dog
lazily chasing next year's game and cim-
lins ignore the hum of TV near their rim
of dream world as visions of birth entice
forth cornfield beans and fertilize them
though grouse and quail must bundle in the ice
as garden seed are bought at January's price.
Although seed catalogs encourage gems
of plants, winter reveries cannot last,
but it's a help to dream tomato stems
to ease discomfort at the winter's blast
and remind us of springs and harvests past.

Who Has Title?

A scenic view of meadow and cowbells
includes brindle leaning over bluebells
to green scum jelling a three-day mudhole,
a beauty unable to kindle joy
in the inspective gaze of the blind mole.
The grubby gentleman can but enjoy

worms and other delights too sweet to tell
to those who envy him in his dark hell.
Eocene vistas gave this bugbear
a rapid pulse beneath a coat of fur
which keeps him warm but hidden from air
in his dark lair away from the owner's
barking, nosing mutt—a notable cur
who seeks under grasses for old mole's fur.
Those ridges in the owner's lawn cause heartburn
and hunting at the hardware for a mole trap
to settle an ancient ownership scrap
that no old, enfranchised mole can discern.

Intersections on the Inland Waterway

Our minds travel ages in mysterious ways
creating images of the past to grace our present
so that our ancestors come alive in our days,
reincarnated in our suns' ascent.
My inland waterway meanders from coastal waters
to mountain streams around the world—
travels that span all continents and only loiter
when my dreams relive my joys and perils.

Fleeing westward, eyes on the smoky ridge
Appalachia raises beyond the coastal plain
a pilgrim seeks relief from city forms that bridge
bright streams until they groan and complain.
Mountains spur the mind of the dreamer
ruing Eden's fall to phallic embrace

yet raise ancestors roasting meatier
portions with hot embers to their taste.

Bodies bound, the leapings of dreams address
fiery circles frying mammoth chops for hikers yelling their hunger. Famished, tearing flesh, *erectus*
grabs and gnaws at the morsels.
Transported to the jungles of New Guinea
I join a jovial crew motoring down the Sepik River
to see what fearsome headhunters used to be
as they enact old ways adorned in marsupial fur.

Next, Nantahalas shed their colored leaves
into the fogs of Cullowhee, Valley of Lilies,
now falling to chimneys and walls and eaves
burned in the fireplace of memories.
Leaving the mountain festival fall
I return to the swamp woods of my youth in flatlands
of Chesapeake where the katydids call
a welcome back to waterways of Powhatan.

There randy tunes of the fish crows and jaybirds crack
the stillness and blend with the hot hum of mosquitoes
and the whistle of the Southern down the track
as I meld with my forebears in time's fast flow.

Ode to a Power Shortage

While the grunt of a dozer
puts bread on our tables

and sculpts a highwall
to the gods of progress,
the advocates of comfortable disease
supply a press
release defying all our disbelief.

We sit and contemplate
the valley hid with fill,
the wrecks of mountains,
the wraith of dirt in disarray
dumped anywhere the Monied please
who borrow from the trees
of yesterday.

What is Mt. Rushmore
other than a wall, they say,
of sculpted dirt and stone, a way
of honoring our past?
So pay tribute to the present need
whose sores we justify to feed
our deficits of energy.

Take It Off! Strip!

When I decry bulldozers tearing the land
you tell me, "Beauty is a biscuit!"
and threaten me with loss of livelihood
and deep breathing in my phone
at three a. m. In the daylight you demand
I see mountains as lumps of mud. You make pits

in serpentine circles where peaks stood
as continental backbone.

On every truck I read the boast, "I dig
Coal!" but where are the signs that add, "with care"?
Yeah, you've made thirty million in five years
of earthy rape and pillage:
so this black gold will be dug with your rigs
you'll turn mountains into empty air
where a man can't live and high walls rear
up to mark a dead village.

In Appalachia it's hard to argue jobs
won't last as long as scars upon the hills
and the scars within the soul that set
our children's teeth on edge
when they look at mountains made into stobs
or leveled into the hollows as fill
to make a fast buck without a regret
for mechanical wreckage.

Silver Lawns

Some green vines turn brown
at the first frost.
It silvers grass, astounds
vegetable stalks
until their sap is lost
as life defaults.
The Cumberlands burn
in October sun,

a brush painting fern
and green trees golden.
Potatoes newly dug
lie on the ground. We cough
as we shiver and shrug
this beauty off.
Rumours spread the word
that wooly worms are furred
so old man Mullins 'sees
bad storms and knows
there'll be an early freeze
and autumn snow.
Leaves that were burnt red
a week ago
crackle under the tread
of boots that go
to meet November wind.
Hiking noisily home tonight
toward the autumn's end,
crunching up the hollow's light
before the hills turn white
to cover tree leaves shed
as snow softly embraces earth,
we seek for new birth
on a squeaky bed.

Namibian Feasts:
If You Want to Save Some Wildlife

Any African ecotourist
should have seen a zebra or a gnu,

a bat-eared fox or an oryx
and perhaps an elephant or two.

But some ecotourists are gourmets
who prize their wildlife cooked.
Ranchers who make their wildlife pay
sell tourists steaks, after they've looked.

Strange beasts abound at Roy's place
where Roy's wife cooks gourmet feasts
for Namib ecotourists' taste—
exotic herbs and gamey meats.

After you've birded Swakopmund
and pelicanned at Walvis Bay
you can drink good beer and eat a ton
of crocodile at Sussusvlei.

If you want to save world wildlife
from people whose habitat has beasts,
then take up your fork and knife.
First look, and scope, enjoy, then eat.

Internal Blues

No Second Flights

My dove of youth that last year flew away
did not return this birthday.
This morning as I stand before the mirror
I see gray age my world abhors.
Tying my halter the old way
I see a rising sun glare at my back today.

My wrinkled face looks foul.
So many times we put on the wrong cowl
for defense against the looming storm
yet can only mourn its harm.
I sent the wrong birds to scout. My birds
could only count the costs I had incurred.

I should have sent a raven—
they wear the proper clothing,
can croak and soar with gutteral ease
and roust those fears that lurk dark seas.
I shall pretend my messengers were dark
and hope to find repose in my mind's ark.

Retrospection

I could not admire the passions I have had
passed down to me from some old stone age man;
although I could not judge them good or bad,
I know their anger has made me sad
and dragged me to that artless, unplanned
but frightened life which embodies glands
that slew old Adam. Doctors brought a cart
to wheel me off before I fell apart.
Age has brought calmer nerves and self-control,
so time has been a friend and enemy
whose patient work has left me quiet and whole
with pastel pleasures found in memories.

Satellite Map

This chart for my too unlit night
is not for Argus eyes imperative
upon the plains of platitudes;
this map I draw with seer-like sight
beyond the streams whose bulls must drive
in forthright cows' beatitudes—
a plat to guide my frequent flights
avoiding bony, starving beeves,
escaping guard of Io eve,
who may betray my latitudes.

A World of Afternoon

Artificers of imagined worlds
are not confined by stored heat—
we can create demesnes with concrete
rules set by imagination alone.

The mind's morning shadows drift away
on sliding cotton and red satin pleats
flouncing red through the day's doorway,
until noon steps through on hot feet
from which white shoes have been pulled to greet
afternoon rainbows whirling in the sun
like drunken tumblers that touch the hair of a girl
who pulls back strands to produce a curl
of lambent hair whose slow recoil's begun
to light flambeaux in the eyes of a hungry fox
who unfolds the colder husk of cloth stocks
past brown shoulders and pink nipples
to steal ivory grapes whose wine he tipples.

In the afternoon sun lizards
and foxes may share the heat in rocks;
we jump charge life with energetic shocks
when mind-created worlds collide.

After Parting

I think my heart a martyr or a saddened bobolink
with serenades of indigo,
so many yesterdays ago

my enemy cut us a pen apart with oceanic sink.
"I will shed tears, " I said, "if you go
because I am not free
to stow
away and follow
you to the cafes of *Paris*."
That must have been ... how many years ago
was it that I had to sew
my love together with the needled foe
of wait 'til I return? An old poet missed his love so:
"I saw you last, I know,
twenty or thirty years
before the clouds grumbled
and the promised flood tumbled
down upon old Noah's ears."
Come back and see the mountainsides
we climbed to watch the sun
burn life into the April myths homespun
shaking their dainty latticed hides
at the disappearing snow on hillsides
foretelling that blooms upon the budding trees
will unite my love and me.

Ragged Tread

From my labored start
faintly remembered
in kaleidoscopic moments
of the motor's whir—
become a part
of horizon's fading remnant:

with studs to clutch
the mud grips of my mind
engage to touch
the ice events
of memory—I grind.

Intentions

Footsteps fell upon an empty path
whose brown leaves betrayed a blight
and a halfway house of harvest chaff
where the crop ends in barn light,
unwinnowed sheaf for a staff.

Dirty feet wandered down a dusty path
indented with uncharted holes
even in dry autumn aftermath
when the heat of the forest doles
last life careless of the hunter's shaft.

I went to harvest delicious fruits
from the orchard ground where apples fell
and bent my ears to hear the owl's hoot
as I considered the sulphurous smell
on my face and hands and boots.

Cover-up

Sometimes my snow melts. My defenses
fade on mountains in my provinces.

As the ruler, I taste the guilt
for deaths that spring from indecision
and building walls of weak confusion.
In land of ruse where weak walls are built
stands the ruler's moldy house of unclear
mind. No children play except in fear
at shuttered windows and cracked sashes
which let in food for the owls.
Mice run under the moon's leer
on old, wind-scattered ashes
while new snows fall as the winds howl
and snow drfts gather in a white cowl.

Requiem for Nick

The tassel-crumpled corn suggests a plan,
a hoped-for harvest of sun-ripened grain,
not barren brush cut down on ditch banks
before it swells with fertility's rain.
Our hands outstretched, we mourn thwarted design
in small relief. Standing on harvest fields
with winter birds we sing decay's dark signs
whose shorter stalks imply our wraith-like deals.
For our farewells we shadows must be brief
for mourning those who haste away from us,
who, like green forests, succumb to a thief
with frosty hands. We must remember thus
that maple buds will redden limbs in light
of spring when love and lilies flash their white.

Seekers

Climbing the Matterhorn
in the morning dawn
means only that we'll wish
by evening that the Hindu Kush
had been the object of our scorn.

Heat Wave Past ...

We fell in love with an ancient notion
of chivalry romancing vows exchanged
beneath an arbor greened by emotion—
lovers' flowery bowers trouveres arranged
with centuries of songs regaling moons.
We woke to find us strangers who too soon
found out, flowers withered, lovers weep
vegetable tears for bouquets in June
as warblers sing. We remembered out of tune
the twilight songs begun with lovers' sleep:
we sit like statues in an antique shop
whose disdain seems apt to signal complete
the wreckage of our vows we made to stop
the living deaths of lovers lacking heat.

Winter Mulch

Snow falls tonight
and I am sad for loss of autumn leaves.
Remember how we sat

and watched leaves fall in clear light?
We were at peace
as the gold and red relief
erased the dark mat
with painted leaf.
I wish that snowflakes
brought quiet without heartache.

Whispers

When I was fifteen
We brought maples and oaks
From the ditch bank and planted our yard
With these and some mail-order trees,
Pecans and peaches and pears green
To the gaze and shading the folks
Throughout the years that they stood guard
And cradled a cool breeze.

Voices murmur at me
From the rooms my hands have emptied
To fill a rented moving truck
For hauling off the furnishings of their lives,
And mine, across the hills and away from the sea
In a nervous excitement of a free
Flight from the thought of being struck
When the mockingbird dives.

Growth

For long, long years I rumbled,
ranted and raved incessantly
about life's shallow worth;

my mad motors grumbled,
and axes were ground by me,
simply to satisfy my dearth;

and then, how odd, humbled;
I realized the soul's great majesty,
increased my girth.

The End of Youth

In dreams I traveled far around
the world with Sinbad
and put aground
on lands that had
not once been bound
by human minds, I thought, since
beautiful Scheherazade
breathed her storied incense
about the Caliph's harsh charade.

Then I journeyed past
tropical isles and slipped
by horizons in a fast
and white-rigged ship
that never returned

to its brilliant slip
after I spurned
the sails that wafted me where
I floated on breezes of purple air

until three wrathful monkeys
summoned dragons and grinned a wry
grin as firedrakes burned my skies
despite my naive pleas.

Debriefing

Conquer, will we ever? Well I doubt it
when a primate urge reigns in my head
and tingles all the nerve-ends that I own
with cries to seek that momentary high
adrenalined with heat to make us writhe
in a damp bed where we awake the dead.

Nerve ends frozen, what's the gain to have them
rendered calm through chilly subterfuge,
the ancient thrill electric over now?
Batteries of DNA always run down,
but human hearts demand a jumper start
and cannot live for long without some heat.

Philemon to Baucis

We have spent long years together
in good times and foul weather.

Our love is not a storybook romance
nor the heat inside the groin
although you know you stir me there,
yet not with a come-hither glance
or memory of the bright dawns
before we settled our affair.
How can an ardent lover say,
"You make me comfortable
and tolerate my follies."
These words do not display
the bellowing rut of a bull
nor kisses by magnolia trees,
but you and I live every day at ease.

Bus Trip

Around a bend a winding of horn,
as my love in green goes riding,
her beauty unseen but gliding
—a waif greeting the dawn.
Inside the station a girl with Chinese eyes
sips her green tea whose color
held in two hands implies
an attempt to do honor
to a most ancient Ming,
but the old Emperors Ching
didn't chew tomato sandwiches away
nor get asked what they would have today
by a short-order Mother Hubbard twin
who is very helpful but hardly Confucian.
She's extremely cheery,

"Don't miss your bus, Dearie.
Don't despair. You're going somewhere."
We're throwing off old traces
and really going places.
Good fortune cookies await us there.
Good-bye, good-bye, my friends,
soon our journey ends,
but my love in green will go riding on,
gliding into the dawn.

Command Performance

Seize the unshaped eager nights for graces
and care that a too-swift clock rushes
us untaught to command our naked feet
and thoughts that prompt our too-quick races
and a sweat on our brows that brushes
off silky clouds surrounding our retreat
with thoughts to give our dappled traces
a bright dawn with sated satyrs' horn
and transfixes our hearts with nagging fears
of losing scenes that touch our eyes and ears
with an eager mouth whose kisses seem born
in lilac odors tasting of sweet tears—
until our batteries revive desire
as artists of life find much love's required.

Afternoon Mail

Yearning for Ponce de Leon's fountain

of youth in moments of dark thoughts that stain
my mind after my latest mailbox trip,
I ponder life's recent rejection slip.
We grant the very old and very young
the right to vent anger that touches hearts
and spills the dark distress that has stung
their minds, but mature age craves silent arts.
In black moments of self-pitying care
I opt for dumb reverie of despair.

Fire Escape

Hunters on the veldt in the Pleistocene—
a group of clever, biped animals
cringe and snarl at night their hate of unseen
fears and snatch embers to brandish fireballs.
Habilis to *Homo sapiens* ran
when fiery hands threw out flame brands.
Today, snarling in kraals, modern sun-hued beasts
with metallic claws sport being man
using high tech to torch savannah lands
with powder-dark hands whose fire's released
when angry hands throw out firebombs.
*Homo sapien*s to terrorist succumbs
as dark genes reassert ancient demands
and new horrors stalk in modern lands.

The Game Is Fixed

Here we must sit, the dealers and the dealt.
Poor poker players have no place to go
and dealers show no love for those who've felt
they could bet weak whores against a foe
who holds a full house hand and bluffs as well,
but we players cannot leave this game and go
to other tables. Our strength cannot impel
a change of games. We play a harsh high-low.

My First Trip to Chicago. 1956

In Chicago I went on a visit
to a very religious family keen
to burn out blacks who had bought a lot
and built a house down their street.
Their eyes expected my eyes complicit
atavism to echo what I had seen
when a young man's rage fueled hatred's rot:
my conscience dredged up a dark conceit.
Growing up in Virginia as a boy
who wanted his buddies to accept him
I acted out sick body's hatred
of *niggahs* with ritual.
Driving down the highway with angry joy
of malt liquor in our hands and a grim
curse on our lips, we sought victims to shred
our tender, civilized shawl.

Laughter erupted in our car as black
bodies jumped from the edge of the road
to escape the rush of our oncoming car
lunging at defenselessness.
As we swerved to avoid the ditch and crack
up, we established ourselves in the code
as we hurled our empty cans as far
as we could in our excess
of cheerfulness. We cheered when we scored a hit
on the figures crouching in the drainage ditch
like a frenzied crowd cheering the home team's
annihilation of their foes.
No matter that the morning after fit
my conscience ill. The emotional pitch
achieved by unleashing our anxious dreams
nourished diseased egos.

After reading Faulkner, I thought men's doubts
had conjured up the hate that threw the cans
and curses. I thought to blame grandfather,
or at least the Klu Klux Klan.
Then I took my trip to Chicago's shout-
filled streets and watched the wary Africans
while they eyed me sidling down their corridor
of concrete jungle land.

Almost alone in a swarm of dark faces
I sensed a hatred akin to what spurred my arm
those nights along the ancient southern way.
I saw myself reflected
in their eyes. I had disowned their race

as not akin to me. The wind from the storm
that shook me blew a black man's hat my way
as forceful gusts directed
a deed of reparation for old guilt.

Stooping down I picked up the brown fedora,
brushed out a bruise, and extended the hat
to a bewildered man,
adjusting the blade of his fear with a tilt;
he stood awhile in an amazed aurora
before extending a hand to take the hat
and absolve my lack of tan.

Thermostat Control

A mystic madman stands in the half-light
at open doors and listens to the dark bird
who sings to announce his evening flight;
the listener hears as a sightless person stirred
to cast off the iron mask whose prison
covers with rusty wrap the mental jumps
he does not trust because his indecision
negates faith in his crimson heat pump.
The angelus of the hermit thrush
in his ears signals his need to retreat
to a place where the sounds of day are hushed
and the sensuous fare of Phaedria's feast
does not lead to the dirty morning daze
for those who lurch within her stupored maze.

Her minions gasp for air like fish upon land.
Burning in the sun the bloating pygmies
cavort with glee upon her floating island
that moves aimlessly to and fro too free,
from gloom to glow—itinerary they feel
will avoid all the rocky, rougher roads
and help to ease their fears with prayer wheels
to provide a fleshy salvation mode:
"Oh for a cup of gladness and delight,"
implore those seeking lotus without life.
To the mystic in his innermost flights
comes a vision of blood beneath a bier
where a butchered bull must feel knife's bite
to make red flow on the proselyte's fears.

The pilgrim's unbroached pump with valves working
lets him see the flight of birds without fear;
now his mental wheels are interlocking
and his higher pressure lets him play the seer.
He can climb to the city on the hill
or take the way of meditation meant
for those who would pure thought distil
under a Bo tree of enlightenment.
Those who repair the interlocking wheels
defeat Phaedria's power to deceive
those whose crimson pulsating machines feel
the strength of their red heat pump to achieve
a victory for soul bird's evening show.
That stirs strong flow in the pilgrim's dynamo.

Comic Blues

Cheers!

Too grand a view of courage, kid,
will mar your chances for success.
Where can a fellow find an Id
these days with nothing to confess?

You say you will face bravely met,
the world, yet seek a trace of mirth.
Let us then, you and me, go wet
our stomachs down and seek new birth.

In Venus' bar no doubt will be
a girl to give us strength to chase
our cares into a Daiquiri.
Let us curse fate bravely, but with grace.

Cakes and Ale

Malvolio, you've had it. Your time has fled
by clock and fire and water, wind, and see

how little dint you made in immorality
by washing mud from social scenes
and whimpering at the foaming head
sucked off the beer and making public preen
to gain a second's immortality
while lovers laugh and lip upon the green
and pay no heed to those who preach obscene.

A Consideration

The Legalists told Mencius
that man is fuss and lust;
he should not seek
for men both good and sleek.
Mencius pondered about
it. He answered in a polite shout:
man is mostly good!
The sage has not been understood.

Street Scene

There walks a man ruing late
down the street tandem with mate,
with sighs and grimaces
marching by his magistrate.

He tends the tares he sowed
early by accepted code;
he now is firmly led to places
where formerly he strode.

He slyly watches calves career
eliciting his secret leer—
subversive in his traces,
responds submissively, "Yes, dear."

Vamp

There goes a pixie prancing
dressed in red,
fitted for summer dancing
on green beds.
She is the glow of evening,
its soft light,
meeting the dark with scheming
looking to the night.
Lithe dance and whispered chants
make sweet moments
while revel girl plants
thoughts in forment—.
whacked hair in boyish cut,
laughing leers pert
upon a row of gleaming teeth,
portrait of a flirt.

Clancy to the Poet

Don't hand me that guff about artist stuff,
just pay your bill and be decent enough.
That's all I ask of any man:
eat well, pay reg'lar, be part of the clan.

You have the God-given right to overstuff.
Disregard the flimflams of fancy.
Its freedom is risky, it's too damn chancy.
Why should you argue the dictates
of pooh-poohing critical pates
when you can dine out with Mrs. Clancy?
Don't starve in a jail, or up in a garret;
art is not food or a cigarette.
Why dally around with a Muse
when you can have any girl you choose,
and gallons of beer and cellars of claret?

Snake-eyes

A philosophy of *carpe diem*
appealed to Abou-hem;
one might say his whim
was the stem of Abou-hem's desire.
So he touched his lyre,
looked forward every day to dine
and sip good wine,
until one unfortunate night,
pushed over a precipice while throwing dice
the hedonist was caught in a trice.
Aware of his plight,
he began muttering prayers;
he repeated them thrice
as the way to Paradise,
but the gist of his impasse
and his unlucky cast
forced him down to devilish affairs.

Adamite Orchardists

He hungered for my apple trees
because his fruits weren't good enough,
but instead of planting other trees
he cut my orchard down to the duff.
His apples win all the prizes now;
he foresees blue ribbons at State Fair
and has ordered a suit for his bow.
He will curse those trees I picked bare.

Indigestion

Metallic cones give off peals
inviting hollow dreams:
our Gargantua's plight
comes from guts of steel
outbelching smoke and screams
to heckle day and night.
It would be fit fuel to inveigh
greatly offended laughter
from old master Rabelais,
whose gross methane after
dinner came from healthy farts,
not from mechanical arts.

Politically Incorrect Advice

In arguments with wives about the shop-
ping men should do, a clever ploy might

be to cite a bit of ancient history:
in Athens women worked indoors, adopt-
ing tasks set forth by Homer as the right
of goddess, wife, and daughter; slavery
must fetch the water jugs; discreetly
they flirted and gossiped at fountain sites,
but matrons cared for children, nursed the sick
and sewed and cooked; for Platea's siege the fight-
ing men were left behind women to sop
the soldiers' hunger. Surely manly tricks
can win our sexual battles if we're slick.
If trained, our wives might even shop for mops.

A Fact of Life

One sunny green-shod day in a very natural way
I climbed out into the world, or you might say fate hurled
me past life's gate across a threshold from warm-
ness out to cold,
I'm told, for I must admit I think I came a bit
unwillingly from my mother's fold.
We passed an hour or two with song
before I stood
upon my feet. I could
not wait for long before I thought me strong
enough to face harsh life. Alas, though Mother paid my fee,
I found the doctor's knife had never cut me free.

Busy Old Fool and Friend

Holding his two heads between his hands
gingerly pleading a hedonist's case,
he addressed his unwelcome visitor,
"You've brought a beautiful day, my man,
but why don't you go away. It's a waste.
I'll stagger back to bed, inquisitor."
To his grief, that morning his bosom friend
had climbed up fast and full of fun to grin
on bloodshot eyes, on yawning, cursing man
who acted out his anger unrehearsed.
Sleepy, shaking his fist with bleak chagrin
he told his buddy, "I don't need a tan."
Eyeing the bottle on his desk, he cursed
his happy friend. Barely able to thin
the gray confusion in his throbbing head,
he addressed his sun, "O jinn of the gin,
I am inclined to think I'm almost dead.
After long nights with Gilby's and a whore
(here he was forced to blink) you are a bore.
You, bragging, bumptious sun, are no fun.
Do *not* force yourself on my attention."
The hotshot denied the plea to vacate.
In his meddling style, he heated away
in disregard of his bosom buddy's pate.
Unhindered by clouds, he blazed out all day.

A Flamboyant's Birthday

For a year Ralph Teutsch has burned among us.
His fiery zeal has seared the sullen flesh
of many a wayward youth in calculus.

Now has his flame burned two and twenty years
toward his last problem in the finite math,
and his presence warms us as it sears.

As geometric wheels turn on our wires
and our machines must calculate his day,
we thank him his bright fuel to our fires.

Eddie Guest, I Know You

You're a poet singing lasses,
sad souls, and sentimental asses.
You've gnawed the bone of sentiment
and whined or wagged your way to print—
would build a house beside a road
and be a friend to man, a mode
of life that moves your readers' hearts,
for they adore your canine arts.
Their handkerchiefs always at hand
to wipe the tear you make to stand
on grim cheeks, they find your aphorisms
turn too blank glass to prisms
that change their earth from dross to gold,
I'm told, that will not tarnish nor grow old.
You coat old truths with stickiness;
you're a mutt with craftiness.

You Can't Hike in Kruger, and the Zulu Are on Strike

The Kruger warden carried on
though the Zulu were on strike.
All night the Zulu drums and Zebu horns
argued Zulu wages should be hiked.

The warden sighed...
"Tsk, tsk!" the warden's lady cried,
"Here come the ecotourists. They can't hike
and the Zulu are on strike.
What *shall* we do?"
"We'll roll out a van or two
and take them to the nearest hide,
let them see a zebra, nyala, or gnu,
or perhaps a lion pride."

"Oh no, dear, haven't you heard?
These are Americans hot after birds.
They beat the bush in their own vans,
but they must eat. How can we feed this herd
with the Zulu striking, every man?"

They met us sadly, wringing hands.
The warden warned, "You can't hike in Kruger
and our food supplies are slender
since the Zulu are on strike,
and their drums don't let you sleep at night."

For us ecotourists things seemed rather grim.
We could not hike in Kruger

and the food seemed mighty meager.
We feared we must pack it in,
but dauntless Ian Sinclair led *us*.
He asked for gnu and kudu steaks.
While we birded aboard our bus,
the kudu marinated
as we counted birds and debated
how cold beers ease stomach aches.

At night the cold beer flowed
while steaks sizzled on the grills
and appetites began to grow—and grow.
Result—a gamey gourmet thrill.

So, in Kruger where you cannot hike,
do not let yourself despair
if the Zulu are on strike.
Just ask to eat some kudu fare.

Adaptation

When boys are small,
they are happily inclined,
yet the elders feel
that all
should toil
manfully and grind
approval's seal.
So boys take castor oil
and learn to make
piled mountains out of mole hills

and take, for victory's sake,
a bitter pill.

Ben Ezra's Fraud

I am growing grayer Robert Browning,
and new medical technology does
not help me accept my senses fading
away like old soldiers. Your promise was
the best was yet to be. The sounds, the tastes,
the smells, the sights have all grown less acute.
I feel a pinch of flab about my waist,
see no reports that I am more astute
than I was when my hair was thick and black.
I must confess I still enjoy my life
even though I fear sudden heart attacks
and wonder how to satisfy my wife
despite deep new wrinkles and little hair,
arsenal of pills and calorie care.
Still, our age gives great boon—we love and live,
and I am loath for the alternative.

www.ingramcontent.com/pod-product-compliance
Lightning Source LLC
Chambersburg PA
CBHW071429070526
44578CB00001B/42